Diospyros Species in Thailand: Their Distribution, Fruit Morphology and Uses[1]

Naoki Utsunomiya, Suranant Subhadrabandhu, Keizo Yonemori, Masayoshi Oshida, Shinya Kanzaki, Fumiaki Nakatsubo, and Akira Sugiura

Utsunomiya, Naoki (*Faculty of Agriculture, Kinki University, Nara 631, Japan*), **Suranant Subhadarbandhu** (*Faculty of Agriculture, Kasetsart University, Bangkok 10900, Thailand*), **Keizou Yonemori, Masayosi Oshida, Shinya Kanzaki, Fumiaki Nakatsubo and Akira Sugiura** (*Faculty of Agriculture, Kyoto University, Sakyo-ku, Kyoto 601-01, Japan*). *Diospyros Species in Thailand: Their Ecology, Fruit Morphology and Uses. Economic Botany 52(4): 343–351. 1998.* Diospyros *species distributed widely in Thailand were classified into four ecotypes, according to their habitat; constantly humid area, alternately dry and wet area, mountainous cool area and all area. Some of them inhabit near dwelling areas or in the paddy field in the village. The young fruit is covered with dense pubescence in most species. The size, shape, and color of mature fruit greatly vary greatly with the species. In most species, the mature fruit has a soft pulp and hard skin. The fruit of six species has been used for dying. Four species produce edible fruits, with color and flavor favorable for breeding of edible* Diospyros *species. The fruit of some species contains some chemicals useful as fish poisoning or of medicines, although the active components have not yet been identified. The edible fruit contained many tannin cells, but the fruit used as fish poisoning and medicines had only a few.*

タイ国にはカキ属樹種が広く分布し、それらの自生地から判断して、これらの樹種は年中多湿な地域、乾季と雨季が明瞭な地域、平地全域、高地の冷涼な地域に適応する４つのグループに分類された。いくつかの樹種は農村の家の近くや水田に自生していた。いずれの種においても若い果実は濃いじゅう毛に覆われている。果実の大きさ、形、色は種によって特徴があるが、ほとんどの種の成熟果実では果肉は軟化するが果皮は硬いままである。調査した樹種のうち果実が染色に利用されてきたものが６種あった。４種の果実は生食に用いられ、これらの果実の特色のある色及び良い香りは生食用カキ果実の育種に有用と思われた。果実が魚毒用や薬用に利用される種もあるが、その有効成分の同定はまだ行われていない。生食用果実は多くのタンニン細胞を含んでいたが、魚毒用及び薬用果実にはタンニン細胞がほとんど含まれていなかった。

Key Words: *Diospyros* species; Thailand; ecology; tannin; medicinal uses.

Diospyros, the largest genus of Ebenaceae family, includes about 400 species. Most of them are important tropical forest resources, but have not been used efficiently. On the other hand, *Diospyros* species in the temperate zone (such as *D. kaki* L.f., *D. lotus* L., *D. oleifera* Cheng and *D. virginia* L.) have been used as fruit crops or ornamental trees, or as tannin resources *Diospyros kaki* especially has been cultivated commercially for its fruit in East Asia and the cultivation area is expanding around the world. Ng (1978) inferred that *D. kaki* arose directly from *D. roxburghii* (= *D. glandulosa*). Some tropical *Diospyros* species may be closely related to the temperate species.

[1] Received 5 July 1996; accepted 27 January 1998.

Economic Botany 52(4) pp. 343–351. 1998

In Thailand, 60 *Diospyros* species have been described by Phengklai (1978). Many of them are distributed in the forest throughout the country. Their fine and hard wood is of high value, and the leaves, bark and fruit of some species have been used as medicinal herbs. Several species produce edible fruit. However, information on their ecology, morphology and uses is limited. In Thailand, from 1992 to 1994, we surveyed the distribution of *Diospyros* species, collected the fruit, and gathered information on their uses by interviewing people in the village. In this article, the ecological features, morphological characteristics and tannin cells in the collected fruits and uses of their fruit are described.

Distribution and Ecological Features

The survey areas and the distribution of *Diospyros* species are shown in Fig. 1. Each species has characteristic environmental requirements. The climate of Thailand has rainy and dry seasons, but the rainfall pattern determining the moisture conditions differs with the districts. The *Diospyros* species are roughly classified into four groups as shown in Table 1 according to the climatic conditions of their habitats.

D. confertiflora, D. toposia, D. sumatrana, D. wallichii, D. transitoria, and *D. cauliflora* are distributed in the southern penninsula and eastern area. In these areas, the amount of rainfall is relatively high and the dry period is very short. We found these species at the riverside or near a waterway in the forest. *D. areolata* often grows along the periphery of mangrove forests. Thus these species adapt to constantly humid conditions.

D. castanea, D. ehretioides, D. gracilis, D. variegata, D. mollis, D. montana, and *D. rhodocalyx* are often found in the central and northeast regions where the dry period lasts for about 6 months, indicating that they tolerate drought conditions well. The former four species inhabit low bush forests. *D. montana* was found in small bushes around temples in the village. *D. mollis* were often found in the temple or home gardens. We found *D. rhodocalyx* growing in paddy fields or along the roadside. *Diospyros decandra* and *D. malabarica* var. *siamensis* are distributed throughout the country except for cool mountainous areas. The distribution area of *D. glandulosa* is localized to a high altitude zone which is sometimes affected by frost. This species was never found in the tropical lowland, and is believed to require a low temperature to break bud dormancy.

The flowering starts at the beginning of the rainy season and the fruit ripens before the end of this season in most species in Thailand. This suggests that flower initiation of these species is induced by drought during the dry season. Drought, which plays an important role in the flowering, is a determinative factor for the distribution of tropical fruit trees (Samson 1980). The drought requirement for flower initiation may be one of the factors determining the distribution of *Diospyros* in Thailand. A short dry spell seems to be sufficient for the flowering of the species growing in wet areas. *Diospyros* species growing in the area where wet and dry seasons are distinctly separated require a long drought period for flower initiation. *D. glandulosa* completes the flowering process during the subsequent low temperature period.

Morphological Characteristics of Fruit

The size, shape and color of fruits of *Diospyros* species in Thailand vary with the species (Fig. 2). Each species has its characteristic calyx. The fruit is large in *D. decandra* and small in *D. ehretioides* and *D. confertiflora.* In many species, the young fruit is covered with dense pubescence which becomes thin or disappears as the fruit grows. However, in *D. wallichii* and *D. areolata* the density of pubescence is still high even on mature fruit. When the fruit matures the skin turns yellowish orange in many species. The fruit of *D. decandra* and *D. malabarica* var. *siamensis* has yellow and reddish brown skin, respectively. Red appears on the brownish yellow ground color in *D. ehretioides.* The fruit of *D. mollis* becomes black when ripened on the tree or within a few days after dropping. When the fruit matures, the flesh becomes very soft but the skin is still hard in most species.

The flesh occupies a large portion of the fruit in *D. decandra, D. glandulosa* and *D. areolata.* On the other hand, the fruit of *D. rhodocalyx, D. mollis, D. ehretioides* and *D. wallichii* has a small volume of flesh. The ripe fruit of *D. malabarica* var. *siamensis* has beautiful reddish orange or red flesh. Fruit of some edible species is fragrant. The mature fruit of *D. decandra* has a strong peach-like flavor. The flesh of *D. glandulosa* smells like that of *D. kaki.*

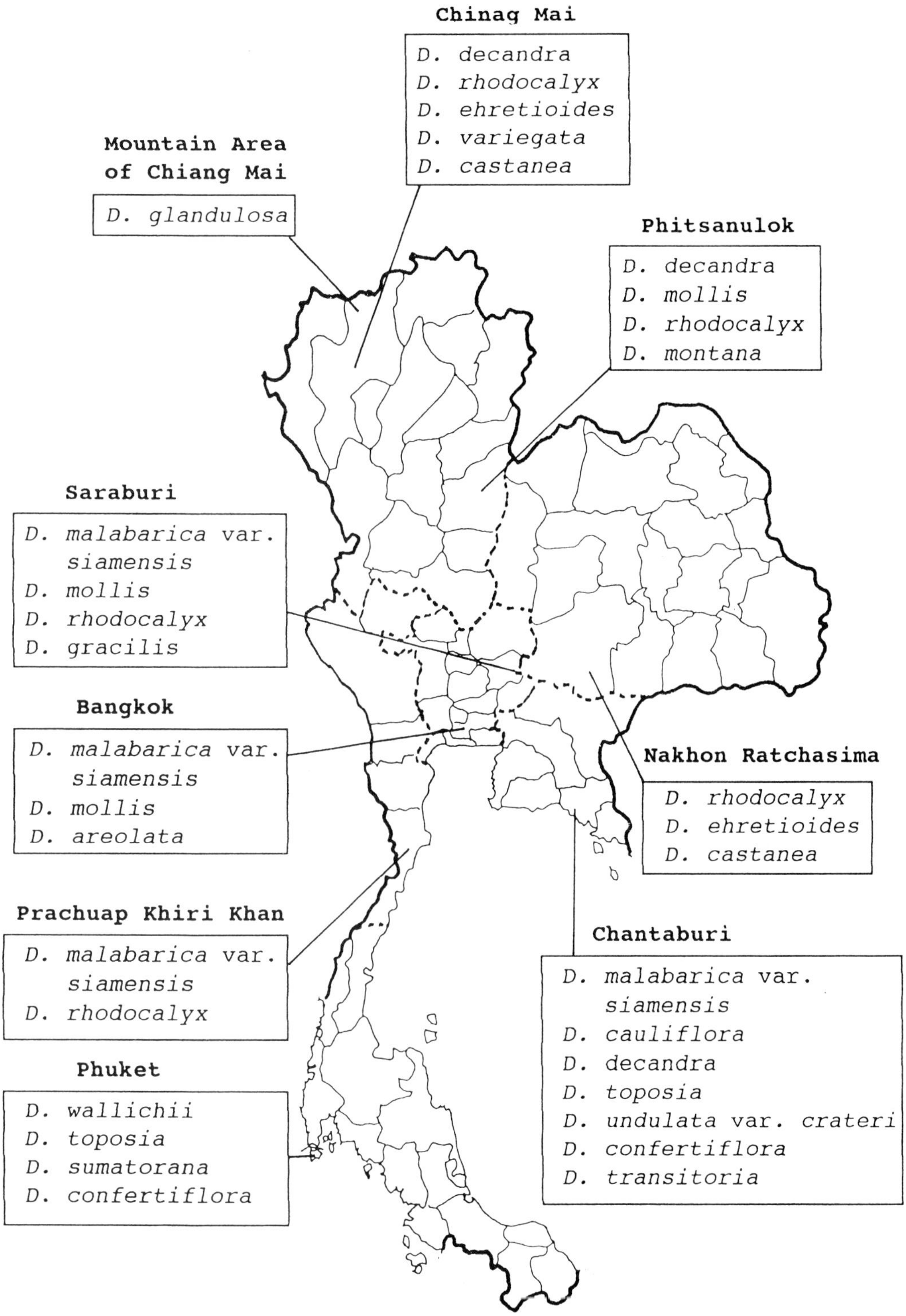

Fig. 1. The survey areas and the distribution of *Diospyros* species in Thailand.

TABLE 1. CLASSIFICATION OF *DIOSPYROS* SPECIES IN THAILAND ACCORDING TO THEIR HABITATS.

Areas	
Wet	*D. confertiflora* (Hiern) Bakh., *D. toposia* Ham., *D. sumatrana* Miq., *D. cauliflora* Bl., *D. wallichii* King & Gamble ex King, *D. transitoria* Bakh., *D. areolata* King & Gamble
Wet and Dry	*D. castanea* (Craib) Fletcher, *D. montana* Roxb., *D. mollis* Griff., *D. rubura* Lec., *D. ehretioides* Wall. ex G. Don, *D. rhodocalyx* Kurz, *D. gracilis* Fletcher, *D. variegata* Kurz
Cool and Wet	*D. glandulosa* Lace
Whole country except for cool area	*D. decandra* Lour., *D. malabarica* (Desv.) Kostel var. *siamenthis* (Hochr.)

The ability of seed formation differs with the species. Many seeds were formed in the fruit of *D. rohdocalyx, D. glandulosa* and *D. mollis.* Seedless fruit was often found in *D. decandra, D. malabarica* var. *siamensis* and *D. areolata.* This indicates that these species may have a great parthenocarpic ability. In *D. decandra* the seedless fruit is called "chan" and the seeded one "Inn" in Thai. The seeds germinated well in most species except *D. wallichii.*

TANNIN CELLS IN THE FRUIT

Existence of tannins in the fruit is one of the characteristics of temperate *Diospyros.* The tannin usually is contained in idioblasts called tannin cells in *D. kaki* and *D. lotus.* The structure and distribution of tannin cells differ with the cultivar in *D. kaki* (Yonemori and Matsushima 1987). However, there is little information on tannin cells in the tropical species of *Diospyros.* Therefore, we compared the tannin cells in the fruit of these tropical species with those of temperate *Diospyros.*

The fruits of *D. wallichii, D. montana, D. mollis, D. ehretioides, D. areolata, D. gracilis, D. glandulosa, D. cauliflora, D. malabarica* var. *siamensis, D. rhodocalyx* and *D. decandra* collected in Thailand were examined. For reference, fruits of *D. kakki, D. oleifera, D. virginia, D. lotus* and *D. rhombifolia* collected from the experimental orchard of Kyoto University, Kyoto, Japan, were used.

To prepare the thin section of mesocarp, a small block of fruit flesh was fixed with 4% paraformaldehyde and 5% glutaraldehyde in 0.05 M phosphate buffer (pH 7.2) and embedded in plastic resin, Acrytron (Mitsubishi Rayon), after dehydration in an ethanol series. Embedded blocks were cut into 2–4 μm sections by a rotary microtome, model 820 (American Optical) equipped with a glass knife. Photomicrographs were taken after staining with 0.05% toluidine blue. The percent area occupied by the tannin cells in the section, and size of tannin and parenchyma cells were calculated by a Macintosh computer (Centris 610) using the public domain NIH Image v.1.55 program after the photomicrographs were scanned by Scan Jet Iilx (Hewlett Packard).

These species were classified into four groups according to the percent area occupied by tannin cells in the section (Fig. 3,4). Interestingly, fruit flesh of *D. rhombifolia, D. montana, D. wallichii* and *D. mollis* contained no tannin cells in the mesocarp. *Diospyros ehretioides* showed a sparse distribution of tannin cells. These species were classified as Group I. Species with the percent area of tannin cells from 10 to 30% were classified into Group II, which includes *D. gracilis, D. oleifera, D. areolata* and *D. kaki, D. glandulosa, D. cauliflora* and *D. virginiana* had tannin cells in 30 to 50% of the area, and were classified into Group III. In *D. decandra, D. lotus, D. rhodocalyx* and *D. malabarica* var. *siamensis,* more than 50% area of the section was occupied by tannin cells and thus they were classified into Group IV (Fig. 4).

The size of tannin cells was much more diverse than that of parenchyma cells (Fig. 5). The size of tannin cells of *D. ehretioides* and *D. decandra* was similar to or slightly larger than that of parenchyma cells, although they were distributed in the mesocarpin the latter species. The tannin cells were much larger than parenchyma cells in other species.

USES

Table 2 shows the uses of various *Diospyros* fruit and their classification. The mature fruit of *D. decandra, D. malabarica* var. *siamensis, D. glandulosa* and *D. rhodocalyx* is edible. They are sweet with some astringency even when fully matured. Although they are familiar to local people, they have not been cultivated commercially. *D. decandra* is often grown as an orna-

Fig. 2. Fruit and leaves of *Diospyros* species in Thailand. **A.** *D. decandra,* **B.** *D. rhodocalyx,* **C.** *D. glandulosa,* **D.** *D. malabarica* var. *siamensis,* **E.** *D. mollis,* **F.** *D. ehretioides,* **G.** *D. montana,* **H.** *D. wallichii.*

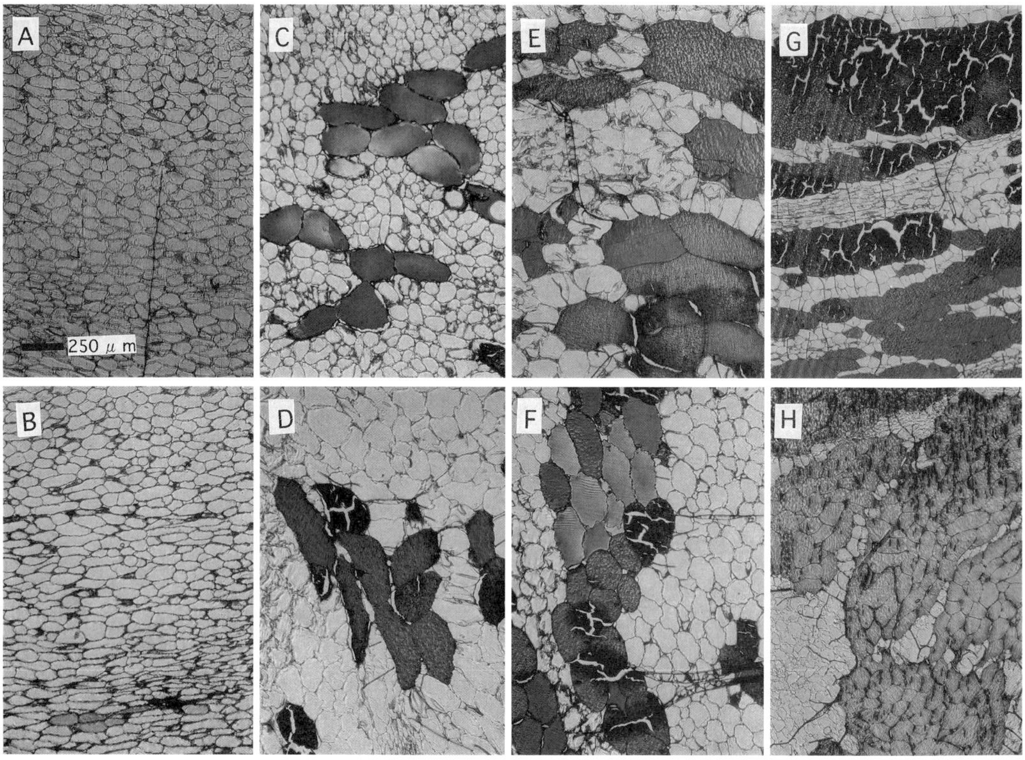

Fig. 3. Tannin cells in the mesocarp of *Diospyros* species. **A.** *D. mollis,* **B.** *D. ehretioides,* **C.** *D. areolata,* **D.** *D. kaki,* **E.** *D. glandulosa,* **F.** *D. virginia,* **G.** *D. decandra.*

mental tree in home gardens or on temple grounds. Local people harvest the fruit of *D. marabalica* var. *siamensis* or *D. rhodocalyx* from trees scattered in paddy fields or growning along the country road side. *D. dasyphylla* Kurz, *D. pyrrhocarpa* Miq. and *D. insidiosa* Bakh are also known to produce edible fruit. However, these species only inhabit the deep forest. We could not obtain this fruit during our survey.

In other tropical areas, some other edible *Diospyros* species have been reported. West Africans harvest the delicious fruit from *D. mespiliforms* trees in Savanna (Campbell 1987). *D. discolor* Willd., indigenous to the Philippines, and *D. nigra* J. F. Gmelin, native to Central America are grown in the Philippines on a small-scale (Coronel 1991; Morton 1987; Ng 1991), but neither of them is cultivated as a marketable commodity.

Because the fruits of tropical edible species have a shape, texture, flavor and taste quite different from those of the temperate species, there is a possibility of breeding new types of persimmons by intercrossing temperate and tropical *Diospyros* species.

In temperate countries like Japan and China, tannin from *D. kaki* and *D. oleifera* has been used for dyeing fishing net or paper, tanning leather or, in some cases, as a proteolytic agent in brewing sake rice wine (Itoo 1980). In Thailand, six species have been used for dyeing (Table 2). The young fruit contains a sticky exudate which is quite astringent.

The poisonous fruits of four species are utilized for fishing (Table 2) as is the young fruit of *D. nigra* (Ng 1991). None of the temperate *Diospyros* species have been used as fish poisons. Only a limited number of tropical *Diospyros* species seem to have fish poisoning substances in the fruit. The poisonous ingredients have not yet been identified, and might not be related to tannin compounds because the fish poisoning species like *D. wallichii, D. ehretioides,* and *D. montana* have no or few tannin cells in the fruit.

Diospyros kaki was originally known to be a

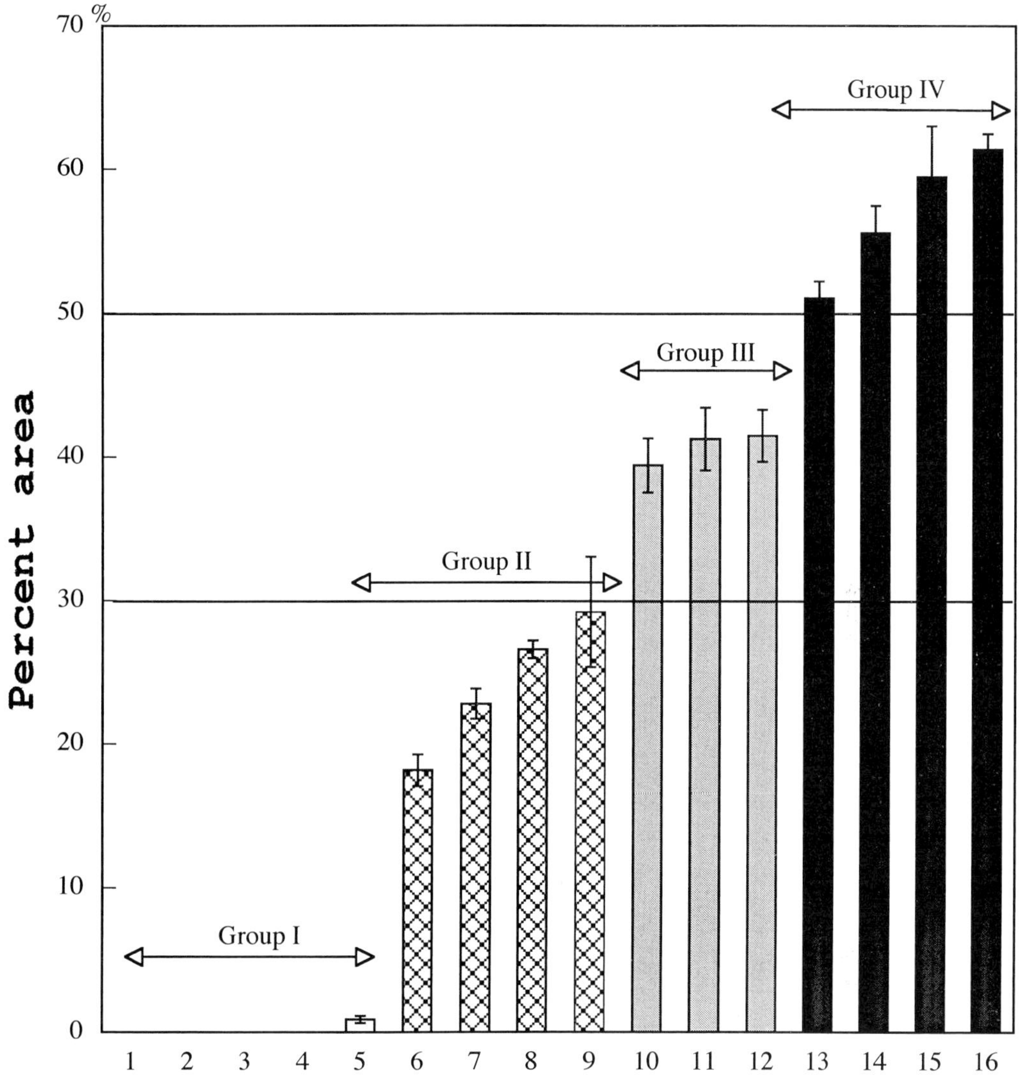

Fig. 4. Percent area occupied by the tannin cells in the mesocarp of *Diospyros* species. Group I: 1) *D. rhombifolia,* 2) *D. mollis,* 3) *D. wallichii,* 4) *D. montana,* 5) *D. ehretioides.* Group II: 6) *D. gracillis,* 7) *D. kaki,* 8) *D. oleifera,* 9) *D. areolata.* Group III: 10) *D. glandulosa,* 11) *D. cauliflora,* 12) *D. verginia.* Group IV: 13) *D. decandra,* 14) *D. lotus,* 15) *D. rhodocalyx,* 16) *D. malabarica* var. *siamensis.*

medicinal plant in East Asia (Flaumenhaft 1982). In our survey, the fruits of only two species, *D. montana* and *D. mollis,* has been used as ethnobotanical medicine in Thailand. Appearance of the two species is similar and the fruit has few tannin cells. The extracts from the fruit of *D. mollis* have been used as an anthelminthic or against tapeworms in the village. The local people informed us that its use requires caution because an excessive dose induces the loss of eyesight. The pharmacological activities in the fruit of these species have been examined (Dhavadee Ponglux 1987; Loder and Mongolsuk 1957; Strum and Zilliken 1972; Yoshihira, Natori, and Kanchanapee 1967). The fruit of *D. montana* is used to treat hiccups, ulcers, urinary diseases, biliousness, dysentery, and bile tract stones. The fruit of *D. decandra* and *D. malabarica* var. *siamensis* is used as an emmenagogue in Vietnam and leucorrhoea in India, respectively. However, such uses have not been reported on either species in Thailand. No in-

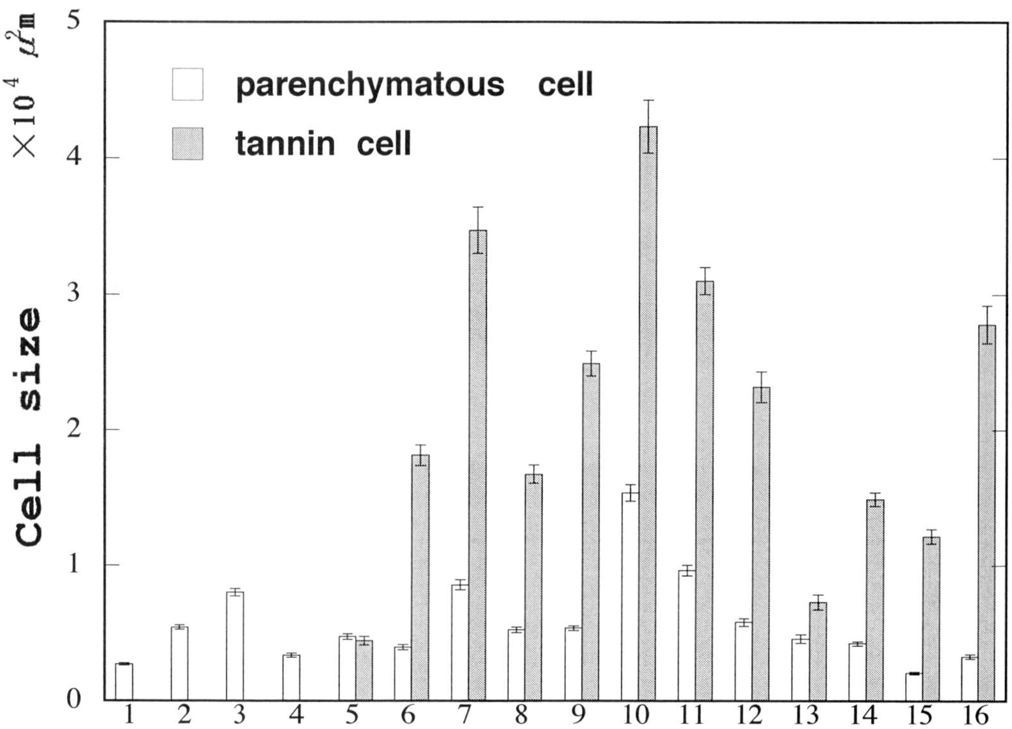

Fig. 5. Size of tannin and parenchymatous cells in the mesocarp of Diospyros species. Group I: 1) *D. rhombifolia,* 2) *D. mollis,* 3) *D. wallichii,* 4) *D. montana,* 5) *D. ehretioides.* Group II: 6) *D. gracillis,* 7) *D. kaki,* 8) *D. oleifera,* 9) *D. areolata.* Group III: 10) *D. glandulosa,* 11) *D. cauliflora,* 12) *D. verginia.* Group IV: 13) *D. decandra,* 14) *D. lotus,* 15) *D. rhodocalyx,* 16) *D. malabarica* var. *siamensis.*

formation on the medicinal use was available on the fruit of other species during this survey. Antifungal activity was found in the fruit of *D. cordifolia* in India (Prasad and Simlot 1982). In Thailand, the leaves, stem, bark, and root of *D. rhodocalyx, D. variegata, D. mollis* and *D. montana* have been used as ethnomedicines. Furthermore, pharmacological activities have been found in these tissues in *D. gracilis* (Tan et al. 1991), *D. wallichii* (Congdon et al. 1981) and *D. sumstrana* (Zakaria et al. 1984).

TABLE 2. USES OF FRUIT OF *DIOSPYROS* SPECIES IN THAILAND.

Uses	Species
Dyeing	*D. malabarica* var. *siamensis, D. cauliflora, D. rhodocalyx, D. areolata, D. gracilis, D. castanea*
Fish poisoning	*D. wallichii, D. ehretioides, D. toposia, D. montana*
Medicine	*D. mollis, D. montana*
Edible	*D. glandulosa, D. malabarica* var. *siamensis, D. decandra, D. rhodocalyx*

CONCLUSION

Diospyros species are important forest resources but environmental requirement for their growth varies among the species in Thailand. Some of them favor a humid condition, others need some dry spells during the growth cycle. On the other hand, *D. decandra* and *D. malabarica* var. *siamensis* can grow under any climatic conditions. *D. glandulosa* can inhabit only in a cool mountainous area. These *Diospyros* species are precious and should be conserved as forest genetic resources. Further studies to clarify the appropriate environmental conditions for their growth, flowering and seed germination are needed.

The uses of *Diospyros* fruits in Thailand are limited. The tannin extracted from the fruit of six species have been used for dyeing. The fruits of four species have fish poison components, which have not yet been identified. Four species, which bear edible fruits, grow naturally near villages, suggesting that their fruits have long been eaten by people. These species have the potential to be improved as horticultural crops, and some may contribute unique fruit characteristics to temperate edible *Diospyros* species. Except for *D. mollis* and *D. montana, Diospyros* fruit has not been used for medicinal purposes in Thailand. Some species, however, may have pharmacologically useful components in the leaves, stems, bark and roots. Further chemical analysis of the fruit will increase the number of *Diospyros* species available for the medicinal uses.

Anatomical studies have shown that the tannin cells in the fruit differ among the species. In our study the edible fruit had many more tannin cells than the species used for fish poisoning and medicine. The analysis of tannin cells will be helpful in clarifying the interspecific relationships in *Diospyros.*

LITERATURE CITED

Campbell, B. M. 1987. The use of wild fruits in Zimbabwe. Economic Botany 41:375–385.

Congdon, G., P. Sirirugsa, V. Lojanapiwatna, and P. Wiriyachitra. 1981. A contribution to the Thai phytochemical survey II. Journal of the Scientific Society of Thailand 7:87–90.

Coronel, R. E. 1991. *D. blanco* A.DC. Pages 151–152 *in* E. W. M. Verheij and R. E. Coronel eds., Plant resources of South East Asia. No. 2: Edible fruits and nuts. Pudoc. Wageningen, Netherland.

Dhavadee Ponglux. 1987. Medical plants. Victory Power Point Corp., Ltd. Bangkok, Thailand.

Flaumenhaft, E., and E. Flaumenhaft. 1982. Asian medical plants in seventeenth century French literature. Economic Botany 36:147–162.

Itoo, S. 1980. Persimmon. Pages 442–468 *in* S. Nagy and P. E. Shaw eds., Tropical and subtropical fruits. AVI Publishing, Inc., Westport, Connecticut, U.S.A.

Loder, J. W., and S. Mongolsuk. 1957. Diospyrol, a constituent of *Diospyros mollis.* Journal of the Chemical Society 1957:2233.

Morton, J. F. 1987. Black sapote. Pages 418–419 *in* Fruits for warm climates. Creative Resources Systems, Winterville, N.C., U.S.A.

Ng, F. S. P. 1978. *Diospyros roxburghii* and the origin of *Diospyros kaki.* Malaysian Forester 41:436–438.

———. 1991. *Diospyros digyana* Jacq. Pages 152–154 *in* E. W. M. Verheij and R. E. Cornel eds., Plant resources of South East Asia. No 2: Edible fruits and nuts. Pudoc. Wageningen, Netherlands.

Panthong, A., W. Ngarmwathana, K. Swatdimongkol, and U. Permphiphat. 1971. Pharmacological evaluation of Thai medicinal plants. Journal of the Medical Association of Thailand 54:490–504.

Phengklai, C. 1978. Ebenaceae of Thailand. Thai Forest Bulletin 11:1–103.

Prasad, B., and M. M. Simlot. 1982. Antifungal activity of fruit of temru (*Diospyros codifolia*). Science and Culture 48:290–291.

Ponglux, D., S. Wongseripipatana, T. Phadungcharoen, N. Ruangrungsri and K. Likitwidhayawut. 1987. Medical Plants. Victory Power point Corp., Ltd. Bangkok, Thailand.

Samson, J. A. 1980. Tropical fruits. Longman, London.

Strum, G., and F. Zilliken. 1972. Constituents of fruit of *Diospyros mollis.* Plant Med. 21:61.

Tan, G. T., J. M. Pezzuto, A. D. Kinghorn, and S. H. Hughes. 1991. Evaluation of natural products as inhibitors of human immunodeficiency virus type 1 (HIV-1) reverse transcriptase. Journal Natural Products 54:143–154.

Yonemori, K., and J. Matsushima. 1987. Morphological characteristics of tannin cells in Japanese persimmon fruit. Journal of the American Society of Horticulture Sciences 112:812–817.

Yoshihira, K., S. Natori, and P. Kanchanapee. 1967. The structure of diospyrol: the principle from the fruit of *Diospyros mollis.* Tetrahedron Lett. 1967: 4857.

Zakira, M., J. A. D. Jeffreys, P. G. Waterman, and S. M. Zhong. 1984. Naphthoquinones and triterpenes from some Asian *Diospyros* species. Phytochemistry 23:1481–1484.

Huron Smith's Ethnobotany of the Hocąk (Winnebago)

Kelly Kindscher and Dana P. Hurlbert[1]

Kelly Kindscher (*Kansas Biological Survey, 2041 Constant Ave., University of Kansas, Lawrence, KS 66047-2906 USA,* k-kindscher@ukans.edu) *and* **Dana P. Hurlbert** (*Department of Systematics and Ecology, University of Kansas, Lawrence, KS 66045 USA)* Huron Smith's Ethnobotany of the Hocąk (Winnebago). *Economic Botany 52(4)352–372. 1998. The Hocąk, commonly known as the Winnebago, are one of the original tribes in the present state of Wisconsin. The field notes of Huron Smith, compiled in the late 1920s and early 1930s, document the extensive use of plant materials by Hocąk people. Smith's notes contain references to 199 vascular plant species in 74 families, with recorded uses for 153 of these species. Medicinal plants (with 117 species) comprise the largest category, followed by food (37 species), and fiber and material uses (22 species). Smith's work is unique for its time because he thoroughly explored the tribal uses of the plants in addition to collecting voucher specimens and photographic plates, and because it remains the most extensive Hocąk ethnobotanical study. Added to Smith's other works of tribes in Wisconsin (Menominee, Meskwaki, Ojibwe, and Potawatomi), the Hocąk ethnobotany broadens the cultural base of his regional compilation of Native North American plant uses. In addition, this is an important body of information for the Hocąk people and those interested in their use of plants.*

Key Words: Hocąk; Winnebago; ethnobotany; Smith, Huron; Wisconsin; North American Indians, Wisconsin ethnobotany.

Hocąk Wazijaci: A Statement

Wąkšik š'ak ra waire šųnų ną. Mą'ųna waną'į cų wąk hahiwi ną. Waną'į na hanąc hik'ų kjawi hesge aire ną. Wažą hįjawi ra, wažą hanąxgųwi ra, goišip hįwįperes hanįhe kjawi hesge ną. Wąkšik wošgą na hirasa aire ną. Waną'į nąkre wažą wokoresge ų ruxuruk wa'ų nąk šąną. Mą hogihi rohą xjį hįjihawi ra, waną'į homąšją hesge ra, hųkurujis hįrahawi ną. Wąkšik hąjąwi na, egi mą hirakara wąk hakererawi ną. že hawa'ų waną'į homąšją že'e hųkurujis hįrahawi ną. Xąwį nąkre, ną nąkre, Hocąk ra hanąc wawiperes hire ną. Jagu hirokų pį wa'ų nąk hišge wawiperes hire ną. Wošgą že hanįwi mąnegi hąjijewire ją ną. Wošgą že'e šge hoxawanį rahe ną. Jąnąkira šge mąką mąšją ra wawiperes hire, hesge wagax nąkre ų hįgikarahere ra, nącge stak jį hąho wawiege, nųnįge, jagu šišik hųgiwawi ra, wąkšik wošgą hį'ųwį šųnų na kenį šge Mąhį Xete ra hajire nį, hegų hąkaranį hąjawi ną. Te pį, wąkšik hijane nąkre hota šge hoit'era gixawanįne, wošgą na šge hirasa.

Wąksik ra, te, Huron Smith Project ąkre ų hįgijirairawi ra, hižąki šąną wa'įginąp šąną. Kansas Biological Survey Team ra racąk pį wakik'ųne wagax te'e hirušją hire ra. Huron Smith Ethnobotany of the Hocąk ra, epa, wagax horucgus hanįwį na eja nįge woiš'ak eja nąžį kšąnąhe ną.

Hocąk Wazijaci: A Statement

Hocąk means "Sacred Language, Parent Tongue, Master Language"

Wazijaci means "Hocąk who lives in pine forests"

One of the teachings of the Hocąk people is that the Creator of all things has given us a mind. We are told to utilize it to the fullest, as far as we can, using all of our senses. We have to remember and recall everything we have heard and seen. This was a way of life with our people. The mind has capabilities beyond anyone's imagination of what it could and can do. Throughout the many centuries we have gradually lost this trait. This degeneration of the most powerful energy source we have has happened due to unrestrained invasion causing genocide of a race of people that were caretakers of this continent. The Hocąk people knew about the gifts

[1] Received 14 July 1997; accepted 10 February 1998.

of plants and their uses and have successfully been a part of their existence since the beginning of this creation. Again due to the elimination of cultural values once practiced by our people there are but a handful of our people who still know of these powerful medicines. It is with a very heavy heart that I have consented to be part of this project. However, regardless of all tribulations we have experienced, we still to this day maintain all of our ceremonies that we have been blessed with, before the appearance of white people, which is very unique since many tribes have lost their languages and ceremonials.

I want to express my personal gratitude to my fellow tribes people who had the foresight to agree to become involved with the Huron Smith Project. The staff of the Kansas Biological Survey team from the University of Kansas, Lawrence, is to be commended for furnishing this invaluable piece of work. The Huron Smith Ethnobotany of the Hocąk (Winnebago) will always have an honored place in our tribal library for the benefit of our people now and in the future. (Kenneth Funmaker, Hocąk Wazijaci Language and Culture Program)

The Hocąk Ethnobotany: History

The Hocąk (Winnebago) are an important tribe in the Great Lakes Region, with a long history in the present state of Wisconsin. Their long occupancy of the Lake Winnebago-Fox River area is supported by historical and archaeological sources (Dorsey and Radin 1910; Lawson 1906; Lurie 1978; Peske 1971) and by the tribe's traditions. Detailed references in their folk literature to the flora, fauna and landforms of the area confirm that Hocąk knowledge of the local flora is extensive (Bigony 1982; Radin 1990).

Huron Smith's previously unpublished draft manuscript "Ethnobotany of the Winnebago" is the only known work that focuses solely on the knowledge and use of plants by the Hocąk tribe. Melvin Gilmore's *Uses of Plants by the Indians of the Missouri River Region* (1991), completed in 1914, contains Hocąk names for 48 medicinal plants used by the tribe on their Nebraska reservation. Andros (1883) listed 23 plants (without Hocąk names) used by the Winnebago and Dakota for medicine during the period when the majority of the Hocąk were being relocated to Iowa, Minnesota, South Dakota, and Nebraska. Radin's extensive work on Hocąk culture in Nebraska, completed in 1923, lists 50 Hocąk plant names but includes no Latin binomials and only limited information on uses (Radin 1990:69–70). Smith's ethnobotany is unique because of its thoroughness (153 useful species, 91 photographic plates, and 201 voucher specimens) and because it is the most extensive research conducted in the original Hocąk homelands in Wisconsin.

How this Manuscript Originated

This project is of mutual interest to botanists and to the Hocąk people, whose elders generously provided the original data to Huron Smith. As Curator of Botany at the Milwaukee Public Museum from 1917 to 1933, Smith had already studied the uses of plants among the Menominee, Meskwaki, Ojibwe, and Forest Potawatomi when he began to work with the Hocąk. Smith spent most of the summer of 1928 based at the farm of a Hocąk family near Wisconsin Rapids, Wisconsin. During that summer, Smith learned about Hocąk culture and plant uses from over fifteen Hocąk men and women. His consultants, acknowledged in his field notes and photographs and in his publication "Among the Winnebago" (Smith 1928b), included the following people: John Bear, the family of Dave Dekorah, Henry Dick, "Cahabįge" (Annabelle Eagle), Mr. and Mrs. Jim Henniger, Mr. and Mrs. Fred Mallory, "Iwahonąžįka" and Beaver Woman (Mr. and Mrs. George Monegar), Mr. and Mrs. Frank Smith, Ray White, and Mr. and Mrs. Ulysses A. White.

Smith's field notes for the "Ethnobotany of the Winnebago" were archived by the Milwaukee Public Museum following his death. The manuscript has been photocopied and circulated among Hocąk people as well as among linguists researching Native American languages. However, Smith's notes were never published or examined in depth by ethnobotanists. The increasing interest in ethnobotany has generated interest in publication of the manuscript. In order to update this manuscript accurately, we consulted the Hocąk Wazijaci Language and Culture Program to obtain the correct Hocąk plant names corresponding to Smith's phonetics and to help us with historical information.

The Hocąk Nation

In 1994 the Wisconsin Winnebago adopted a new constitution, which officially changed their name to the Hocąk Sovereign Nation. Hocąk (or

HoChunk), the name they have always called themselves, means "People of the Big Voice" or "Sacred Language" (Hocąk Nation Department of Historical Preservation 1996). Traditions generally place the origin of the Hocąk at a place called Red Banks, located on Green Bay (Lurie 1960, 1978; Radin 1990:3, 17). Stories of the Hocąk and related tribes tell that over time, groups of Hocąk people moved away, becoming the Oto, Iowa, Missouri, Dakota, and several other tribes (Lurie 1978; Radin 1990:2–4). The name "Winnebago," used for the Hocąk tribe, is of Algonquian origin, while "Sioux," used for the Hocąk family of tribes and languages, is a French derivation of an Algonquian name. Other interpretations of names for the Hocąk are given by Dorsey and Radin (1910) and Lurie (1978).

At the time of earliest European contact, the Hocąk lived in the area of Green Bay, the Fox River, and Lake Winnebago in east central Wisconsin (Lawson 1906; Lurie 1960, 1978; Dorsey and Radin 1910; Thwaites 1902). They were surrounded by central Algonquian neighbors, the Menominee, Potawatomi, Sauk, and Meskwaki (Fox). The Menominee were also one of the original tribes inhabiting the Green Bay region, while the Potawatomi, Sauk, and Meskwaki were driven from the eastern Great Lakes region into Wisconsin by pressure from the Iroquois (Dorsey and Radin 1910; Lurie 1978).

Early European contacts with the Hocąk were made by Jean Nicollet in 1634, and by other French explorers, fur traders and priests (Lawson 1906; Lurie 1960, 1978; Thwaites 1902). By 1670, the initially numerous and powerful Hocąk were severely reduced in numbers by warfare with the Illinois, their neighbors to the south (Lurie 1960; Radin 1990:5–17), and by epidemics, which occurred in the area as early as 1666 (Kay 1984). The Hocąk and neighboring tribes participated in the fur trade from approximately 1665 to 1840. Intensified use of an expanded resource base during this period was associated with increases in tribal populations (Kay 1984) and dispersal into smaller settlements (Lurie 1960, 1978).

White encroachment on Hocąk lands began in the 1820s when armed bands of prospectors intruded into Hocąk lands where lead could be mined (Abbott 1988). Pressure by lead miners, lumbermen, and settlers for access to Hocąk land forced the tribe to accept a series of cession treaties with the U.S. government from 1825–1837. The Hocąk were thereby deprived of their lands east of the Mississippi (Fay 1966; Lurie 1978; Smith and Carstensen 1974). The 1837 treaty, signed under duress by a delegation that had no authority to cede lands, was considered to be fraudulent by the Hocąk (Lurie 1966, 1978; Merrell 1876). Most of the Hocąk were moved to the Turkey River in Iowa in 1840 (De La Ronde 1876; Merry and Green 1989; Thwaites 1892), and thereafter to three different locations in Minnesota (1846–55), to South Dakota (1862), and finally to Nebraska, where a reservation was given by treaty in 1865 (Lurie 1978). From 1838 until 1874, the Wisconsin Hocąk led a fugitive existence in the face of repeated attempts by the U.S. Government to remove them (De La Ronde 1876; Lurie 1978; Thwaites 1892). About six hundred Hocąk people took up homesteads under special legislation enacted in 1881 (Lurie 1978), leaving them scattered throughout eleven Wisconsin counties (Smith 1928b). Today, all Wisconsin Hocąk lands have been repurchased. The 4,700 members of the Wisconsin Hocąk Nation hold title to 2,000 acres of land (Hocąk Nation Department of Historical Preservation 1996).

The Hocąk originally had access to a wide range of ecological communities and a rich and varied resource base, as their homelands are within the ecotone or floristic tension zone between the prairie-forest province and the northern hardwoods province (Curtis 1959:15–20). The archaeological sites associated with Hocąk culture were located in the Fox River drainage basin, which flows into Green Bay (Peske 1971). This area was noted for its highly productive fishery and wild rice beds (Jenks 1900; Thwaites 1902). European-Americans who explored the Fox River waterway described its "channel choked with fields of wild rice" and "myriads of ducks and waterfowl" (Schoolcraft 1975). In addition to these riverine-lacustrine resources, the Hocąk "were in easy reach of prairies, oak openings, southern hardwoods, northern hardwoods, and conifer stands" (Peske 1971). The Hocąk maintained villages and agricultural lands near waterways, but spent months away from the villages hunting, fishing, and gathering. Hocąk food resources thus included cultivated corn, beans, squash, and tobacco, the Great Lakes fisheries, wild rice, and waterfowl, and abundant game and wild berries (Bigony 1982;

Fig. 1. Huron Smith in herbarium with ethnobotanical specimens. (Milwaukee Public Museum staff, date unknown.)

Funmaker 1983; Radin 1990:61–70; Thwaites 1902).

It is likely that the Hocąk relied heavily on wild plant and animal resources during the period of their attempted removal from Wisconsin. However, by the time Huron Smith was working with the tribe, he portrayed the Hocąk people as agricultural, raising "potatoes, corn, beans, carrots, hay, grain, hogs and chickens" (Smith 1928b). The Hocąk people at this time based their economy on seasonal work, picking berries, selling handicrafts, harvesting crops, and hunting and trapping (Lurie 1961:8–17, 1987). Smith did not attempt to quantify the relative importance in the Hocąk economy of cultivated versus wild foods, or the degree to which traditional materials were still used for crafts and items for household use.

Huron Smith

Huron H. Smith (Figure 1) was Curator of Botany at the Milwaukee Public Museum from 1917 to 1933 (Barrett 1933a). Beginning in 1921, he began investigating the uses of plants by Wisconsin Indian tribes. Smith attributed his initial interest in ethnobotany to being called upon frequently "to identify plants or parts of plants used by various Indian tribes" (Smith 1923). Before his untimely death at age 49 in an automobile accident, Smith had published three of a proposed series of six ethnobotanies of Wisconsin tribes. These included the ethnobotany of the Menominee (Smith 1923), the Meswaki (Smith 1928a), and the Ojibwe (Smith 1932). The ethnobotany of the Forest Potawatomi was published later in the year of his death (Smith 1933).

Smith understood that "the Winnebago were the original inhabitants of Wisconsin, at least so far as the period immediately preceding White occupation is concerned" (Smith 1933). He mentions their cultural and linguistic differences from the adjacent Algonquian tribes as part of his interest in working with the Hocąk (Smith 1928b). Smith had completed fieldwork for the Hocąk ethnobotany and mentioned this in a footnote in his Ojibwe study (Smith 1932). In

Barrett's published eulogy he reported the following:

> [Smith's] last day of life had been devoted to writing on his Winnebago paper and he had carried on this work up to within three hours prior to his death, leaving the notes and data in place upon his desk where they could be taken up immediately upon his return. As is inevitable in the handling of such original observations and data, it will be extremely difficult for anyone else to take up these notes and complete the work. It is hoped that this can be done later but for the present it will be necessary to hold it in abeyance (Barrett 1933a).

Smith was unusually suited to his work as an ethnobotanist in the early part of the century. He understood the need to combine botanical fieldwork with ethnology and linguistics. Smith was adept at acquiring Native languages and often spent his summers with indigenous peoples (Barrett 1933b), learning their language while paying specific attention to plant names. He believed that "the botanist must sharpen his ethnological observing faculties" and he stated that "the investigator must be able to write the Indian names so that they can be understood again when pronounced to other Indians of the same tribe, as a basis of checking back on the truth of the information" (Smith 1928a). He believed it was important to spend time in the local communities because different people knew different species of plants, and deplored the brevity of time that he was spending with each tribe (Smith 1932). He recognized that Native knowledge of plant use was being lost by elders dying and through cultural change (Smith 1933).

Smith's interest in indigenous culture and religion extended to both the traditional Medicine Lodge and the Native American Church, which many Hocąk embraced in the early 1900s (Hill 1990; Radin 1990). He published a version of the Medicine Lodge origin story and a Hocąk translation of the Lord's Prayer (Smith 1928b). In his last published work, Smith wrote: "Previous experience with other tribes has taught the writer that the way to an Indian's confidence is to recognize his philosophy of life, to treat sacredly the things he holds sacred and to practice hospitality and generosity in dealing with them" (Smith 1933). Smith gave liberal donations of tobacco and other gifts to people with whom he worked (Smith 1921) and developed friendships with members of several tribes. He noted that a Native person "is quick to appreciate favors and to acknowledge the respect that is given to him by the White man, and becomes confident when he realizes that his confidence is not abused" (Smith 1932). From his personal letters (archived at the Milwaukee Public Museum) we found that he attended ceremonies and powwows and called the 1928 Hocąk pow-wow he attended a "very good sight."

Although Smith's notes and writings reflect some of the cultural biases of the times, he did not believe that Whites were culturally superior to Indians. Smith was disturbed that there was not more respect for the indigenous way of life and wrote that "it is the fallacy of the white man in trying to impose his culture on other peoples and in always assuming that it is superior to any other way of living" (Smith 1932). He recognized with regret that the U.S. government was involved in deliberately discouraging treatment by medicine men, and that through its educational efforts was also discouraging the Medicine Lodge ceremonies (Smith 1932).

Overall, Smith greatly appreciated the wisdom of the people he came to know. While working with Native people, Smith made great effort to collect the plant species that he found in the area where tribal members lived. He would then ask the Menominee, Meskwaki, Ojibwe, or Hocąk people what names and uses they knew for the plants he collected. Smith believed that many of the medicinal plants could be important for medical purposes and stated:

> Much of the knowledge of white men originated from studying the Indian plant uses, in the early days. Eclectic practicioners sought the Indian herbs and observed carefully what parts of the plant were used. The mass of early information was sifted scientifically by the students of medicine, and finally tested physiologically on animals. Perhaps sixty-five per cent of their remedies were found to be potent and are included in our pharmacopeias; the other thirty-five per cent were discovered to be valueless medicinally (Smith 1932).

Smith's keen insights and warm appreciation of traditional knowledge helped make his work a valuable contribution to our body of Native American ethnobotanical knowledge.

Methods

Huron Smith's Data Collection

Huron Smith's methods for gathering ethnobotanical data among the Hocąk were similar to

Fig. 2. Huron Smith's camp at the White farm, Wisconsin Rapids, Wisconsin. (Huron Smith, 1928.)

the methods used in his earlier ethnobotanical studies of Native peoples of Wisconsin (Smith 1923, 1928a, 1932, 1933). During most of the summer of 1928 Smith camped at the Ulysses A. White family farm, 10.5 km south of Wisconsin Rapids, Wisconsin (Figure 2), where he began learning the Hocąk language, met the people, and participated in community life (Smith 1928b).

From his notes, it appears that Smith collected plants and ethnobotanical information simultaneously. An informant's name frequently appears at the start of a series of specimens, and each entry identifies both the species and its uses. Given that Smith participated in aspects of his hosts' daily and ceremonial life, it is likely that he used participant observation techniques; however, as was common for his time, Smith did not describe his interview techniques in detail.

The Milwaukee Public Museum's Department of Anthropology has archived the glass plate negatives for 91 of Smith's photographs. The plates depict his Hocąk work, including consultants and their families, ceremonies, processing of plant materials (Figures 3–6), and food.

Voucher Specimens

Smith collected 201 herbarium specimens representing 190 species during the course of his work with the Hocąk. Smith's voucher specimens are housed in the Herbarium of the Botany Department of the Milwaukee Public Museum (Smith 9167—9367 MIL). He reported uses for another 14 species for which he collected no specimens. As in his earlier works, Smith collected species for which no use was identified by the Native people he consulted, believing that further work with other consultants was likely

Fig. 3. Mrs. George Monegar shelling sweet corn. Wood County, Wisconsin. (Huron Smith, 1928.)

to reveal additional uses. He used *Gray's New Manual of Botany,* 7th ed. (1908) for plant identification and nomenclature.

Methods for Manuscript Completion

After learning of Smith's Hocąk manuscript, we contacted the Milwaukee Public Museum and offered to complete and update it. We typed and edited a photocopy of the handwritten manuscript, highlighting words and phrases in need of possible correction. In November 1994,

Fig. 4. George Monegar washing *Acorus americanus* rhizomes for winter use. Wood County, Wisconsin. (Huron Smith, 1928.)

Fig. 5. George Monegar and granddaughter, Fanny White, cleaning supply of *Acorus americanus* rhizomes for drying. Wood County, Wisconsin. (Huron Smith, 1928.)

Fig. 6. String of *Acorus americanus* rhizomes as George Monegar hung them to dry. Wood County, Wisconsin. (Huron Smith, 1928.)

we visited the Museum and compared our typewritten text with Smith's original field notes. We also examined all of Smith's other published and unpublished works, as well as his correspondence. This process enabled us to verify all but a few illegible words of Smith's notes, and to place this work in the context of his previously published ethnobotanies.

Voucher Specimen Verification and Nomenclature

During our November 1994 visit to the Milwaukee Public Museum, we examined Smith's plant voucher specimens and verified plant identification, in collaboration from the Museum's Department of Botany. Vascular plant identification was verified according to Gleason and Cronquist (1991). All scientific names have been updated to current nomenclature for vascular plants (Kartesz 1996), mosses (Anderson 1990; Anderson, Crum, and Buck 1990) and fungi (Smith, Smith, and Weber 1981).

Current Hocąk Linguistic Information

The linguistic information and phonetic key Smith used for the Hocąk language came from at least two Hocąk informants. John Bear spent the winter of 1927–28 at the Museum recording stories with anthropology curator W. C. McKern, and Fred Mallory compiled a Hocąk syllabary for Smith (Smith 1928b).

The Hocąk Wazijaci Language and Culture Program was established in 1993 to maintain the viability of the Hocąk language and its cultural significance. The Program provided the Hocąk plant names corresponding to Smith's phonetics. This work places Smith's Hocąk plant names in conformity with currently accepted spelling and phonetic symbols. We did not attempt to verify correspondence of the Hocąk name to the botanical name, as that would require a fluent Hocąk-speaking botanist.

Hocąk Plant Uses

The Hocąk ethnobotany contains 199 vascular plant species in 74 families, one alga, two mosses, and two fungi. Smith recorded uses for 153 of these species (Appendix 1). The largest use category is medicinal plants with 117 species, followed by food and beverages (37 species) and fiber and material uses (22 species). The vascular plant families with the largest number of useful species are the Asteraceae (22 species), followed by the Rosaceae (12), Fabaceae (10), Pinaceae (6), Ericaceae (5), and Salicaceae (5).

Smith's notes on Hocąk plant use contain specimen numbers, Hocąk and English common names, and Latin binomials, followed by the uses of the plant. To bring the notes into a format which could be compared to his other works, we have organized the plant data by family within use categories (medicine, food, dye, fiber and material uses, and animal medicines and foods). Although the nomenclature has been updated, we have also retained Smith's nomenclature to facilitate comparison with his earlier works.

We have edited the manuscript minimally for clarity, by expanding abbreviations, adding articles, and correcting minor grammatical errors. Our comments and interpretations within the plant list are bracketed. Through our study of his other manuscripts, we feel certain that Smith would have edited the text further. However, we wanted to retain the original flavor of his vernacular speech and to avoid altering his meaning by our own interpretations.

Medicinal Plants

Smith recognized "the sanctity of . . . medicinal knowledge" of Native people and observed in his earlier publications that medicines must be gathered using the proper method (Smith 1923, 1928a, 1932, 1933). However, he wrote only that the "Winnebago observe the same precautions in gathering plants and place as great stress upon the manner of gathering as do other Wisconsin Indians" (Smith 1928b), without mentioning the religious observances related to the gathering, preparation, and administration of medicinal plants. Hocąk culture specifies that a medicine's efficacy depends on appropriate payment for the privilege of learning about it (Lurie 1961:61–65). Because this information is not included in his field notes, we do not know whether Smith ever learned the full details of medicinal plant use. He may have omitted specific methods for plant gathering and use out of deference to his consultants, or perhaps he disregarded this information as botanically irrelevant.

Smith noted with interest that Native people sometimes cultivated medicinal plants (e.g., see *Rumex orbiculatus* below). *Acorus americanus* and *Gymnocladus dioicus* are two other medicinal species in the Hocąk ethnobotany which probably were cultivated or propagated by indigenous peoples (Curtis 1959:463).

Plant Foods and Beverages

The Hocąk traditionally relied on both cultivated and wild foods. Gardens around old villages were said to be "as large as the distance covered when you shoot an arrow three times" (Hocąk Nation Department of Historical Preservation 1996). Smith took interest in his Hocąk collaborators' industriousness in agriculture and noted that they raised a variety of crops and livestock (Smith 1928b). His field notes describe Fred Mallory as having "several garden patches, raising potatoes, carrots, beans, peas and other garden truck." The field notes contain lists of names of garden vegetables, including eight varieties of corn and seven varieties of beans, which are not reproduced here. Despite this interest in agricultural production, Smith affirmed the value of wild foods, noting that "Indian foods are always described by the Winnebago as being [healthful] for the partaker."

Before European contact, Native peoples were actively involved in managing plant communities and landscapes through the use of fire (Curtis 1959:461–462). Blueberries, mentioned below as an important food, were encouraged by burning. State restriction of burning underbrush resulted in a decline of this important wild crop (Lurie 1961:116). White ownership of cranberry bogs also removed this wild resource from Native control (Lurie 1978). Nevertheless, at the time of Smith's work with the Hocąk, food gathering was still important. Food gathering activities described by Lurie (1961) include a reference to stealing wild beans from mice (see *Amphicarpaea bracteata* below) and gathering water lily roots. Smith does not discuss methods of food preparation or preservation; however, some of these methods are described in Lurie (1961: 8–14) and Radin (1990:70).

Appendix 1

Hocąk Ethnobotany, Huron Smith's Notes

Smith's notes are organized within use categories (medicinal plants, foods and beverages, dye plants, fiber and material uses, animal uses), following the format he used in his other published works. Each use category contains plants arranged alphabetically by family and species. Each entry is organized as follows: Botanical name (Smith's original identification), common name [standard common name], collection number, "Hocąk name" (Smith's translation), and uses and cultural information. Plant entries are numbered and are cross-referenced to other use categories (M=medicinal; F=food or beverage; D=dye; C=fiber and material culture; A=animal food or medicine). All collection numbers are Huron H. Smith's. Plants are native to the region unless otherwise noted.

Hocąk Medicinal Plants

Several of the plant species included in this section are toxic. Medicinal plants should not be experimented with or used without proper observances and guidance.

Vascular Plants

Acoraceae, Sweet Flag Family

1. *Acorus americanus* (Raf.) Raf. *(A. calamus* L.) Calomel [Calamus], Sweet Flag. Smith 9358. "Mąkątek" (Medicine bitter) or "Mąkąkerekerep" (Medicine spotted). A very important medicine for physic and colds, but very dangerous unless only a little bit is used. It is great medicine for camping out. Spray it around the tent to keep out spiders and snakes.

Anacardiaceae, Cashew Family

2. *Rhus hirta* (L.) Sudworth (*R. typhina* L.) Staghorn Sumac. Smith 9171. "Haznįhu" (Berries, water leaking out). The green leaves at the top are cooked to cure stomach-ache and diarrhoea and cramps. It is also a stomachic to clean out the system. The inner bark and the root bark are both used for poultices for sores. The berries are also used in combinations with other medicines. (F120, C163)
3. *Toxicodendron radicans* (L.) Kuntze (*Rhus toxicodendron* L.) Poison Ivy. Smith 9340. The root and leaves are used. Chop a little very fine and open up a boil or carbuncle and put a little in, and it will draw the swelling. One must know just how or else it is very dangerous.

Apiaceae, Carrot Family

4. *Cicuta maculata* L. (*Conium maculatum* L.) [Common water-hemlock.] Smith 9218. "Mąnųske" (Smells good). The roots are used to restore lost appetite.
[Both *Cicuta* and *Conium* are considered to be highly toxic and should not be experimented with. It is possible that the original contributor may have confused *Cicuta* with *Sium* L., water-parsnip.]
5. *Heracleum maximum* Bartr. (*H. lanatum* Michx.) Cow parsnip. Smith 9309. "Mąką'apxete" (Medicine big leaves). The roots are used for the sweat bath. The Hocąk sweat bath has a three foot pit over which a small wigwam is erected . . . The covering is burlap or flour sacks. In this pit, the kettle or flat pan of water is placed and the hot rocks [are placed] in on top of the medicines.
6. *Sanicula marilandica* L. Black snakeroot. Smith 9313. "Mąką'apzazac" (Medicine leaves fine small roots). Chew the root up, spray it on the feet and no snakes will approach you.

Apocynaceae, Dogbane Family

7. *Apocynum androsaemifolium* L. Spreading dogbane. Smith 9192. "Waruhaxąwį" (Fireflies' weed). The leaves are used as a medicine for babies when they have the colic. (C164)

Aquifoliaceae, Holly Family

8. *Nemopanthus mucronatus* (L.) Loes. Northern holly. Smith 9235. "Mąkąhas" or "Nąco" (Stick green). When someone wants to vomit, they use the bark for tea.

Araceae, Arum Family

9. *Arisaema triphyllum* (L.) Schott. [Jack-in-the-pulpit.] [No specimen.] Root is "Waxge." The root is made into a bitter compound much like a mustard plaster for neuralgia or rheumatism.

Araliaceae, Ginseng Family

10. *Aralia nudicaulis* L. Sarsaparilla. Smith 9276. "Mąkąkirikiri" (Medicine soft). When [one is] cut with a sharp instrument, the root is cut and mashed to make a poultice to cure.
11. *Aralia racemosa* L. Indian spikenard. Smith 9180. "Pejąhu" (Sand hill crane-lake). The root is boiled for sores, boils and carbuncles.

Aristolochiaceae, Birthwort Family

12. *Asarum canadense* L. [Wild ginger.] [No specimen.] "Wamąxe." A food seasoner, used to make a tonic tea. (F122)

Asclepiadaceae, Milkweed Family

13. *Asclepias exaltata* L. (*Asclepias phytolaccoides* Pursh) [Tall] milkweed. Smith 9307. "Mąhįc." The root is a medicine. A tiny piece boiled is a lactuary for a squaw [to increase lactation]. (F123, C165)
14. *Asclepias tuberosa* L. Orange milkweed [Butterfly-weed]. Smith 9363. "Mąkąska" (Medicine white). A very great remedy for the Hocąk. The root is chewed and placed in a wound to heal. In case of hemorrhage one should drink it [the root].

Asteraceae, Sunflower Family

15. *Achillea millefolium* L. (*A. lanulosa* Nutt.) Yarrow. Smith 9191. "Mąkąwirirotapanahi" (Medicine smudge). The smoke is an important medicine used to revive consciousness.
16. *Ageratina altissima* (L.) King & H.E. Robins. (*Eupatorium urticaeifolium* Reichard.) White snakeroot. Smith 9285, 9303. "Wakąmąką" (Snake medicine). The root is macerated and used to poultice a rattlesnake bite and cure it. The bite won't swell.
17. *Ambrosia artemisiifolia* L. Ragweed. Smith 9195. "Mąkąhikikuruža" (Medicine headache wash). The tops of this are used to make a wash to cure headache.

18. *Artemisia ludoviciana* Nutt. [White sage.] Smith 9364. "Xąwįskarawirotapąnąhi." A smudge to revive consciousness.
19. *Aster cordifolius* L. var. *sagittifolius* (Wedemeyer *ex* Willd.) A.G. Jones (*A. tradescantii* L.) [Arrow-leaved aster.] Smith 9206. "Poaxų" (Sweat). Used in the sweat bath.
20. *Aster furcatus* Burgess. Forked Aster. Smith 9207. "Poaxų." Used in the sweat bath.
21. *Aster sericeus* Vent. [Western silvery aster.] Smith 9365. "Paxšišik'ųmąką" (Diarrhoea medicine). For colic and diarrhoea, eat the leaves only. Eat them raw, just as they are.
22. *Erechtites hieraciifolia* (L.) Raf. *ex* DC. [Pilewort.] Smith 9200. "Poaxų" (Sweat). Used in the sweat bath to tincture steam with healing ingredients.
23. *Erigeron strigosus* Muhl. *ex* Willd. (*E. ramosus* (Walt.) BSP.) Fleabane. Smith 9204. "Poaxų" (Sweat). Used in the sweat bath.
24. *Eupatorium maculatum* L. (*E. purpureum* var. *maculatum* (L.) Darl.) Joe Pye. Smith 9220. "Wirotapąnąhi"(Smudge). Used as a smudge for illnesses.
25. *Eupatorium perfoliatum* L. Boneset. Smith 9221. "Mąką'apgihap" (Medicine stem three leaves). The tea is used to cause sweating and break up a fever. If one spits blood, it will cure that, too.
26. *Helianthus giganteus* L. [Giant sunflower]. Smith 9225. "Poaxų" (Sneeze). Used for the sweat bath.
27. *Helianthus strumosus* L. [Rough-leaved] sunflower. Smith 9257. "Sawazi" or "Hįnųc" (Yellow legs). Of the class called "Poaxų" or snuff. The fumes of the leaves on coals are used to cure headache.
28. *Liatris scariosa* (L.) Willd. Blazing star; Dotted button snakeroot. Smith 9300, 9302. "Cesįchošok" (Buffalo's tail; like buffalo tails). The root extract cures sunburn. The powdered root is a healing, dusting powder for cancer, sore throat, etc. The root is dried, powdered, and swallowed dry, followed by a drink of water. Swallow down just a little at a time. It is also used for poulticing and tied on over night. (A184)
29. *Oligoneuron rigidum* (L.) Small (*Solidago rigida* L.) [Stiff goldenrod.] Smith 9362. This is much like *S. speciosa* in use. A blood purifier and a great female remedy, taken three times a day.
30. *Prenanthes alba* L. White lettuce. Smith 9238. "Ną'apparas." When a boy goes swimming sometimes he gets lumps all over his body. Then he rubs with this, and they all go away.
31. *Pseudognaphalium obtusifolium* (L.) Hilliard & Burtt (*Gnaphalium polycephalum* Michx.) Many headed everlasting. Smith 9202. "Mąkąwirirotapąnąhi" (Medicine smoke or smudge). When someone is bad sick, use a funnel to smoke and revive them.
32. *Rudbeckia hirta* L. Black-eyed Susan. Smith 9354. "Poaxų." Used for the sweat bath.
33. *Solidago canadensis* L. Canada goldenrod. Smith 9214. "Poaxų" (Sneeze). Used for the sweat bath.
34. *Solidago speciosa* Nutt. [Showy goldenrod.] Smith 9361. "Mąkąrejųserec" (Medicine root long). For incontinent urine. The best blood purifier of the Hocąk.
35. *Solidago uliginosa* Nutt. [Northern bog-goldenrod.] Smith 9222. "Poaxų" (Sneeze). Used for the sweat bath.

Betulaceae, Birch Family

36. *Alnus incana* (L.) Moench. [Speckled] alder. Smith 9234. "Hapuruc" (Bark good to eat). When the stomach is sour and out of whack they eat the bark.

Brassicaceae, Mustard Family

37. *Armoracia rusticana* P.G. Gaertn., B. Mey., & Scherb. (*Radicula armoracia* (L.) Robinson) Horseradish. Smith 9336. "Mąkątakac" (Medicine hot). Used in compounds to cure old sores. Very good for internal cramps. [Naturalized from Europe.]
38. *Brassica nigra* (L.) W.D.J. Koch. Black mustard. [No specimen.] "Xąwįpąnątek" (Medicine with strong smell). The seeds are ground up for tonic and used in compounds for internal ailments, colds, and stomach troubles. [Escaped cultigen, native to Europe.]
39. *Lepidium virginicum* L. Peppergrass. Smith 9188. "Xąwįsurohą" (Lots of seed). When one is tired and the feet and legs ache, this plant is cooked and the tea is used to wash the tired members.

Campanulaceae, Bellflower Family

40. *Campanula aparinoides* Pursh. Long leaf stitchwort, [Marsh-bellflower]. Smith 9244. "Mąką'apserec" (Medicine leaf small long). Used to make a steam to inhale.
41. *Lobelia inflata* L. Indian tobacco. Smith 9258. "Wąkšiktanį" (Indian tobacco). Used for smoking in Indian ceremony. Before the white man came, they used to use it almost exclusively for every day smoking. Does this mean Hocąk had no native tobacco? [George] Monegar's snuff for a cold is *Lobelia inflata* L., slippery elm [*Ulmus rubra* Muhl.] bark, coffee tree [*Gymnocladus dioicus* (L.) K. Koch] bark, and ginseng [*Panax quinquefolius* L.] root, powdered together. **[This species is considered to be quite toxic.]**

Caprifoliaceae, Honeysuckle Family

42. *Diervilla lonicera* P. Mill. Bush honeysuckle. Smith 9187. "Mąkąnąksik" (Medicine—little stick). The root is cooked for a tea to clean out after childbirth. Also used by both sexes to make urine come.
43. *Sambucus nigra* L. ssp. *canadensis* (L.) R. Bolli (*S. canadensis* L.) Elderberry. Smith 9272. "Hicocox" (Hollow stem). A teaspoon of the inner bark in a cup of hot water is a quick physic.
44. *Sambucus racemosa* L. Red elderberry. Smith 9312. "Sosoc." The bark is physic and emetic. It is also used as an injection in constipation.
45. *Triosteum aurantiacum* Bickn. Tinker's weed, [Horse-gentian]. Smith 9308. "Mąkąkirikirik" (Med-

icine soft) or "Mąkąhazmįnąk" (Medicine with berries on). Injections of the tea are given for stomach troubles and chronic constipation, in the regular Indian way. They drink the tea for kidney trouble. One of the greatest Hocąk medicines.

Celastraceae, Staff-tree Family

46. *Celastrus scandens* L. [American bittersweet.] [No specimen.] "Mąkązi" (Medicine yellow). The root is used in compounds, especially the one I drank for a cold. This had nine medicines in it.
47. *Euonymus atropurpurea* Jacq. Wahoo. [No specimen.] "Nąksikhazmįnąk" (Little stick berries). The inner bark and the root bark both were used in treating chills and fevers. The hot tea is used for a foot bath. The root bark is a tonic, and is used very thinly diluted for women to drink during childbirth.

Cornaceae, Dogwood Family

48. *Cornus amomum* P. Mill. [Knob-styled dogwood.] Smith 9288. "Cawaruc" (Deer feed). They smoke the bark.
49. *Cornus racemosa* Lam. (*C. paniculata* L'Hér.) Panicled dogwood, [Northern swamp-dogwood]. Smith 9263. "Mąsigųsge" (Arrow wood). They use the bark for smoking, and it has almost the same flavor but is milder than their regular kinnikinnick. The inner bark and the root bark is medicine as a cleanser for female illnesses. (C168)
50. *Cornus rugosa* Lam. (*C. circinata* L'Hér.) Silky cornel, [Rough-leaved dogwood]. Smith 9255. "Rugįšucge" (Smoking bark). It is used as a kinnikinnick and the bark is efficient in a combination for measles.

Cucurbitaceae, Gourd Family

51. *Cucurbita pepo* L. Squash. [No specimen.] "Wicawacozu." The seeds are used to dispel worms. The shells are removed and the seed eaten raw.
52. *Echinocystis lobata* (Michx.) Torr. & Gray. Squirting cucumber. Smith 9304. "Xą'o'oke" (Hoot owl). The roots are used by some of the Hocąk. The seeds are used as a urinary [medicine].

Cupressaceae, Cypress Family

53. *Thuja occidentalis* L. Arbor vitae. Smith 9306. "Waziparasge" (Pine cedar). One of the greatest Hocąk medicines. Combined with red willow bark and cedar leaves pounded together, it is drunk cold, for curing measles.

Dennstaedtiaceae, Bracken Family

54. *Pteridium aquilinum* (L.) Kuhn (*Pteris aquilina* L.) Bracken fern. Smith 9293."Cošerekehu" (Fern). This root [rhizome] is used in several remedies, most of them for female complaints.

Ericaceae, Heath Family

55. *Vaccinium angustifolium* Ait. (*Vaccinium pensylvanicum* Lam.) Blueberry. Smith 9262. "Hastįk" (Wonderful berries). The berries are dried and sometimes added to medicine to flavor it. This is the type of tonic used to stimulate a lost appetite. (F130)

Euphorbiaceae, Spurge Family

56. *Euphorbia corollata* L. Flowering spurge. Smith 9267, 9360. "Ną'apraxatake" (Leaf milk); "Xąwįsaska." A piece of the root 2 ½ inches long is used to clear out the stomach. The leaves are steeped to make a tea to cure a baby's colic.

Fabaceae, Bean Family

57. *Amorpha canescens* Pursh. Leadplant. Smith 9270, 9331. "Xąwįsku" (Sweet root); "Taxumąką" (Burnt medicine). The leaves are medicine for scalds. It is powdered, then wet and put on. (F133)
58. *Baptisia alba* (L.) Vent. (*B. leucantha* Torr. & Gray) [No specimen.] "Capakginųsge." The root is a single remedy to use for injured womb alone. Cook the root and mash it to form a poultice to bind on. Wash with water and draw out the inflammation. Change twice a day until healed.
59. *Desmodium cuspidatum* (Muhl. *ex* Willd.) DC. ex Loud. (*D. grandiflorum* (Walt.) DC.) Tick trefoil. Smith 9261. "Waraxgaxgapkeparasti" (Stickers flat). The root is used as a seasoner for other medicines. It gives a good flavor, but very little is needed.
60. *Gleditsia triacanthos* L. Honey Locust. Smith 9357. "Nąksikpahįk" (sharp bushes[?]). The root is used in a hot bath or sweat bath.
61. *Gymnocladus dioicus* (L.) K. Koch. Kentucky coffee tree. [No specimen.] "Nąpacaknąkhu." They steep the bark or the root bark and drink it. For colds, take a hot bath, and drink this tea as hot as you can stand. This is one of the ingredients in [George] Monegar's snuff recipe [see *Lobelia inflata* L.]. I took pictures of this. George Monegar was quite anxious to locate where this grew so he could get some. I told him about the Fountain City, Wisconsin grove [located] 1 ½ miles south of town towards Winona, Minnesota. (C170)
62. *Lespedeza capitata* Michx. [Bush-clover.] Smith 9341. "Xąwįzi" (Weed yellow). The root is used in the sweat bath.
63. *Robinia pseudoacacia* L. Black locust. Smith 9356. The root is used in a hot bath or sweat bath.

Fagaceae, Beech Family

64. *Quercus alba* L. White oak. Smith 9169. "Nąhaska" (Tree nuts). The bark and root bark are used in mixture for curing flux. (F135)
65. *Quercus ellipsoidalis* E.J. Hill. Hill's Oak. [Northern pin-oak.] Smith 9280, 9339. "Piksikrakekhu"; "Piksigu'ap." The root bark is a medicine used in compounds. (F136)
66. *Quercus macrocarpa* Michx. (var. *oliviformis* (Michx. f.) Gray) Bur oak. Smith 9316. "Piksigu." The bark is used in combinations of medicines. (F137)
67. *Quercus rubra* L. Red oak. Smith 9170. "Huksi-

gu" or "Piksigu" (Nut tree). The inner bark is mixed with apple bark for worms. (F138, A188)

Hamamelidaceae, the Witch Hazel Family

68. *Hamamelis virginiana* L. Witch hazel. Smith 9177. "Huksik" (Squirrel berries). The root is used as a seasoner for other medicines, to make them taste right.

Iridaceae, Iris Family

69. Iris versicolor L. Blue flag. Smith 9250. "Mąkąsagre" (Medicine fast). A strong physic and a quick one. It is dangerous and must be used correctly, as an overdose would kill. For a baby, use one-half inch; for an adult, use one inch only.

Juglandaceae, Walnut Family

70. *Juglans cinerea* L. Butternut. Smith 9283. "Cazuke" (Fruit tree). The bark is a physic. (F140, D156)
71. *Juglans nigra* L. Black walnut. [no specimen.] The tree itself is "Cakhu." The bark is boiled down with maple sugar until it is about the consistency of black strap molasses, and used for a blood tonic. A teaspoonful in the morning every other day is the dose. (F141, D157)

Lamiaceae, Mint Family

72. *Monarda fistulosa* L. Wild bergamot. Smith 9284. "Poaxų" (Sweat). Used in the sweat bath. They inhale the fumes to cure a cold.
73. *Nepeta cataria* L. Catnip. Smith 9273. "Nįkjąkmąką" (Baby medicine). A sweetened tea is given to babies when they are prone to cry and not sleep. Then they will get a good sleep. [Naturalized from Europe.]
74. *Prunella vulgaris* L. Self heal. Smith 9294. "Wirašarak" (Heal all). The plant itself is used to heal up sores.

Liliaceae, Lily Family

75. *Clintonia borealis* (Ait.) Raf. Northern clintonia. Smith 9260. "Šųmąką" (Dog medicine). The root is used to kill worms in a dog, and in a person, too. (A189)
76. *Maianthemum racemosum* (L.) Link (*Smilacina racemosa* (L.) Desf.) False spikeweed [False Solomon's seal]. Smith 9217. "Wakewaruc" (Coon berries). The root is a physic.

Monotropaceae, Indian Pipe Family

77. *Monotropa uniflora* L. Indian pipe. Smith 9291. "Xąwįska" (Flowers white). This is used as a "Poaxų" or smudge to revive consciousness in one ill who has fainted.

Myricaceae, the Bayberry Family

78. *Comptonia peregrina* (L.) Coult. (*Myrica asplenifolia* L.) Sweet fern. Smith 9184. "Mąkąihira" (Medicine sweat bath). When someone catches cold this is put into a sweat bath to cure him.

Nymphaeaceae, Water-lily Family

79. *Nymphaea odorata* Ait. (*Castalia odorata* (Ait.) Wood) Sweet white water lily. Smith 9241. "Kecoksik" (Red turtle feet). The root is cooked to make a female remedy to cure after a woman gives birth.

Osmundaceae, Royal Fern Family

80. *Osmunda claytoniana* L. Ostrich fern [Interrupted fern]. Smith 9179. "Cųkeresge" or "Cųšereke" (Stand up—lots of leaves). The root is medicine. (C173)

Oxalidaceae, Wood Sorrel Family

81. *Oxalis stricta* L. Yellow sorrel. Smith 9277. "Xąwįsku" (Weed sweet). Give to a baby sick with colic. (F144, D158)

Papaveraceae, Poppy Family

82. *Sanguinaria canadensis* L. Bloodroot. Smith 9367. "Pexhišuc" (Red gourd). This is a dye and a medicine also. (D159)

Pinaceae, Pine Family

83. *Larix laricina* (DuRoi) K. Koch. Tamarack. Smith 9325. "Casak." The gum and the bark are medicines. The gum is used on wounds and cuts. (C174)
84. *Pinus resinosa* Soland. Norway or Red pine. Smith 9168. "Ruğihašarašarake." (Tree has no branches.) The pitch from the cones is mixed with some kind of fat to make a salve to heal sores.
85. *Pinus strobus* L. White pine. Smith 9167. "Wazi." The white pine is used by the Hocąk and was important to them. It is plentiful around in Wood Co., where the soil is sandy and jack pines and Norway pines are second and third in order of occurrence. They use the gum boiled out of the cones to cure sore throat, and also to apply on a sore which would be apt to leave a scar when it heals. Use this and there will be quick healing and no scar left. They also cook the inner bark and make a poultice to cure burns and make them heal quickly. The leaves are also used in the sweat baths. (C176)
86. *Tsuga canadensis* (L.) Carr.. Hemlock. Smith 9355. "Sįhu" (Rice stem). The leaves and bark are used in compounds. It is a medicine for a child. Grind the red part of the bark, and when dry its flavor is good. Eat it when dry to cure diarrhoea.

Plantaginaceae, Plantain Family

87. *Plantago media* L. Plantain. Smith 9323. The leaves heal old sores. They chop the leaves as fine as they can get them for a poultice. [Native of Eurasia.] (A190)

Polygalaceae, Milkwort Family

88. *Polygala sanguinea* L. Milkwort. Smith 9226, 9227. "Cesįcošok" (Buffalo tail). The whole plant is

used for a tea to cure sunburn. If there is cancer in the throat, this is gargled. (A191)

Polygonaceae, Buckwheat Family

89. *Rumex orbiculatus* Gray (*R. britannica* Amer. auct. non L.) [Great water-dock.] [No specimen.] "Hišuc" (Water red). This large root does not occur right here, but the Hocąk can get it at the grist mill at Necedah or at Kilbourn. It is one of their greatest medicines, a great tonic and for consumption. George Monegar didn't have any growing but had seed which he got from the Meskwaki at Tama, Iowa, and was going to plant it to raise his own medicine root. (D160)

Potamogetonaceae, Pondweed Family

90. *Potamogeton natans* L. [Floating pondweed.] Smith 9252. "Kecoksikxųnųnįk" (Turtle feet small water weed). This is much used in combination with other medicines.

Ranunculaceae, Buttercup Family

91. *Anemone cylindrica* Gray. Anemone. Smith 9335. "Mąkąwira'aphi."
They use the fuzz part to chew up and put on boils or carbuncles and bandage. After 24 hours they will open. It is somewhat similar in strength to poison ivy, but is not so dangerous a remedy. The root tea is an eye lotion for sore eyes.
92. *Coptis trifolia* (L.) Salisb. Gold thread. Smith 9320. "Xąwįtanįsara" (Weed club leaves), "Rejųsįnį" (root yellow). The root is used to cure sores in a baby's or an adult's mouth.
93. *Thalictrum dasycarpum* Fisch. & Avé-Lall. Meadow rue. Smith 9281. "Xąwįpąnąpį" (Smells good). Used to make a sachet bag and to perfume smoke.

Rhamnaceae, Buckthorn Family

94. *Ceanothus americanus* L. [New Jersey tea.] Smith 9332. "Cawaruc" (Deer food). The root is fine medicine for women. (A192)
95. *Rhamnus alnifolia* L'Hér. Buckthorn. Smith 9183. "Nąconįk" (Little green stick). The inner bark is a strong physic, and the root too.

Rosaceae, Rose Family

96. *Aronia arbutifolia* (L.) Pers. Chokeberry. Smith 9236. "Tocgiğik." When the berries are dry in the fall, they are ground up and used to stop flux or diarrhoea.
97. *Potentilla norvegica* L. (*P. monspeliensis* L.) [Strawberry-weed.] Smith 9198, 9297. "Hišucxųnųnįk" (Yellow flower). The root is a remedy to clean up the afterbirth.
98. *Prunus pensylvanica* L.f. Pin cherries. Smith 9176. "Nąpakšuc" (Red cherries). The cherries are dried for medicine, pounded in a mortar to a powder and then steeped to make a tea to stop flux.
99. *Prunus serotina* Ehrh. Wild black cherry. Smith 9264. "Nąpakwijanįk" (Tree cherry drunk). This is called the drunk cherry, because if you eat too many you get drunk. The inner bark is used as a seasoner in 20 kinds of combinations, and it sweetens and gives a good taste. A tea of its bark is considered a good tonic. (F148)
100. *Rosa carolina* L. (*R. humilis* Marsh.) Pasture rose. Smith 9196. "Hassuc" (Red berries). The root is cooked for a baby medicine when the mother is going to have another baby. The [first] baby will get sick easily then and needs this remedy to keep it well. The skin of the rose hip is a stomachic. The root of rose is for diarrhoea.
Rosa carolina L. (*R. humilis*), gall on stem. Smith 9345. This gall was gathered on the chance that it might have particular medicinal properties but it did not. [Specimen consists of stem only; authors could not verify to species.]
101. *Rubus allegheniensis* Porter. Blackberry. Smith 9173. "Hassep" (Black berry). The root is used to make a tea to cure sore eyes. (F150)

Salicaceae, Willow Family

102. *Populus grandidentata* Michx. Large tooth aspen. Smith 9174. "Wasge." This bark mixed with red oak was used to cure worms. (C178, A194)
103. *Populus tremuloides* Michx. Trembling aspen. Smith 9175. "Wasge." No recognition of difference from *P. grandidentata* in name, although the Hocąk know that the trees are different. The use is the same. (C179, A195)
104. *Salix amygdaloides* Anderss. [Peach-leaf willow.] Smith 9351. "Ruğixete." The root bark is used in a mixture for soreness and for female weakness.
105. *Salix bebbiana* Sarg. [Beaked willow.] Smith 9318. "Ruki" (Crooked). They say of this one that it never grows straight. They use the root bark in compounds.
106. *Salix candida* Fluegge *ex* Willd. [Sage-leaved] Willow. Smith 9317. "Ruğixoc" (Willow). The root is medicine in compounds.

Sarraceniaceae, Pitcher-plant Family

107. *Sarracenia purpurea* L. Pitcher plant. Smith 9271. "Canącawa" (Deer's ear). The root, though small, is the part used. It is a female remedy to drink when sick at the stomach of pregnancy.

Scrophulariaceae, Figwort Family

108. *Verbascum thapsus* L. Mullein. Smith 9203. "Caskanącawa" (Sheep's ear). Heat the leaves and apply to reduce a swelling. [Naturalized from Europe.]
109. *Veronicastrum virginicum* (L.) Farw. (*Leptandra virginica* (L.) Nutt.) Culver's root. Smith 9359. "Mąkąski" (Medicine root bitter) or "Mąkąrejųzi" (Medicine root yellow). This is a physic, but also is used for a poultice for a pain anywhere.

Smilacaceae, Catbriar family

110. *Smilax tamnoides* L. (*S. hispida* Muhl. ex Torr.) Bristly sarsaparilla, [Bristly greenbriar]. Smith 9292. "Waxacsep" (Black stickers). If the berries are ground up and the powder gets in the eyes, it will blind and no medicine can cure you. The root flavor is esteemed and the root is used in a tonic medicine. (A196)

Tiliaceae, Linden Family

111. *Tilia americana* L. Basswood. Smith 9315. "Hįšge." The root is used for female weakness. (C180)

Ulmaceae, Elm Family

112. *Ulmus rubra* Muhl. (*U. fulva* Michx.) Slippery elm. Smith 9282. "Mąkąrak Ap" (Elm leaves). Used as a sore throat lozenge and in many combinations.

Verbenaceae, Vervain Family

113. *Verbena hastata* L. Blue [common] vervain. Smith 9223. "Mąkąrejųksuksik" (Medicine—fine roots). The fine hair roots are used for female weakness.

114. *Verbena stricta* Vent. Blue [hoary] vervain. Smith 9301. "Caxiwico" (Green flowers). When someone is bad sick, vomits and spits blood, this root stops it.

Vitaceae, Grape Family

115. *Vitis* sp. Wild grape. [No specimen.] "Hapsįc." The vines are sappy and in the spring exude a sap which can be used as a hair tonic. They also hard boil the root in water to get a hair tonic. (F154)

FUNGI

116. *Calvatia gigantea* (Pers.) Lloyd. Giant puffball. [No specimen.] "Wanąğipoaxų" (Ghost sneeze). Used for a haemostatic, but it is too strong by itself and so is mixed with other medicines.

117. *Scleroderma citrinum* Pers. (*Lycoperdon* or *Sclerodermatum*). Smith 9278. "Wanąğipoaxų" (Ghost sneeze). When the nose bleeds, a small quantity is puffed up the nose to stop it.

PLANT FOODS AND BEVERAGES

VASCULAR PLANTS

Aceraceae, Maple Family

118. *Acer rubrum* L. Red maple. Smith 9172. "Wisebu" (Black growing). The Hocąk like the maple syrup made from this, better than the sugar maple. They say it is sweeter and whiter in color than sugar maple. (C162)

Alismataceae, Water-plantain Family

119. *Sagittaria latifolia* Willd. (*S. latifolia* var. *gracilis*) Broadleaved arrow head. Smith 9224. "Sįporo" (Rice round). This is the Indian potato whence St. Paul, Minn. was named. In early times they dried and stored it.

Anacardiaceae, Cashew Family

120. *Rhus hirta* (L.) Sudworth (*R. typhina* L.) Staghorn sumac. Smith 9171. "Haznįhu" (Berries, water leaking out). The fruit is used for a beverage, sweetened by maple sugar. (M2, C163)

Araliaceae, Ginseng Family

121. *Aralia hispida* Vent. Bristly sarsaparilla. Smith 9208. "Tošąnąk Axųcge" (Otter armband). In olden times the root and the berries were used as food. The root is still used in vegetable soup, but the berries are not used now.

Aristolochiaceae, Birthwort Family

122. *Asarum canadense* L. [Wild ginger.] [No specimen.] "Wamąxe." A food seasoner cooked with any hard meat like fish or raccoon to make it tender and to give it good flavor. (M12)

Asclepiadaceae, Milkweed Family

123. *Asclepias exaltata* L. (*A. phytolaccoides* Pursh). [Tall] Milkweed. Smith 9307. "Mąhįc." The flowers or buds are eaten in soups in the spring. (M13, C165)

124. *Asclepias syriaca* L. Milkweed. Smith 9253. "Mąhįc" (Oatmeal). The flowers and buds are used in soup.

Asteraceae, Sunflower Family

125. *Taraxacum officinale* G.H. Weber *ex* Wiggers. Dandelion. Smith 9201. "Xąwįzi" (Make wine). They use the blossoms to make wine when somebody is going to get married. [Native of Eurasia.]

Betulaceae, Birch Family

126. *Betula papyrifera* Marsh. (*Betula alba* L. var. *papyrifera*) Canoe birch. Smith 9185. "Nąhaska" (Tree—white skin). The inner bark is a substitute for white man's tea. (C166)

127. *Corylus americana* Walt. Hazelnut. Smith 9256. "Huksik" (Indian nut). Food. (C167)

Chenopodiaceae, Goosefoot Family

128. *Chenopodium album* L. Goosefoot, Lamb's quarters. Smith 9190. "Raxgemąkejahağep" (Old ground growing). Used as greens when it first comes out, cooked with beef or any meat in soup. [A European weed.]

Ericaceae, Heath Family

129. *Gaultheria procumbens* L. Wintergreen. Smith 9209. "Wašjįkpųc" (Rabbit's nose). Indian beverage tea.

130. *Vaccinium angustifolium* Ait. (*Vaccinium pensylvanicum* Lam.) Blueberry. "Hastįk" (Wonderful berries). Smith 9262. Food. The berries are dried. (M55)

131. *Vaccinium macrocarpon* Ait. Cranberry. Smith 9251. "Hocąke." Where the Hocąk get their name. A food.

132. *Vaccinium myrtilloides* Michx. (*V. canadense* Kalm.) Huckleberry, [Velvetleaf-blueberry]. Smith 9328. "Hawų." Food.

Fabaceae, Bean Family

133. *Amorpha canescens* Pursh. Leadplant. Smith 9270, 9331. "Xąwįsku" (Sweet root); "Taxumąką" (Burnt medicine). The root is used as a food. (M57)
134. *Amphicarpaea bracteata* (L.) Fern. (*A. monoica* (L.) Ell.) Wild bean. Smith 9212, 9314. "Hųnįkboija" (Wild bean; Indian beans; Sets in the ground). The Indians eat the beans and storage tubers. They follow the rat trails and find storage of half a bushel or more. When beans grow full size and are harvested in October they are highly relished by the Hocąk.

Fagaceae, Beech Family

135. *Quercus alba* L. White oak. Smith 9169. "Nąhaska" (Tree nuts). Soup is made from the acorns like hominy is made. (M64)
136. *Quercus ellipsoidalis* E.J. Hill. Hill's oak [Northern pin-oak]. Smith 9280, 9339. "Piksigu'ap"; "Piksikrakekhu." The acorns "Huc" are to eat. (M65)
137. *Quercus macrocarpa* Michx. (var. *oliviformis* (Michx. f.) Gray) Bur oak. Smith 9316. "Piksigu." They eat the acorns. (M66)
138. *Quercus rubra* L. Red oak. Smith 9170. "Huksigu" or Piksigu" (Nut tree). They make soup from the acorns in the same way as above [see *Q. alba*]. (M67, A188)

Grossulariaceae, Gooseberry Family

139. *Ribes cynosbati* L. Prickly gooseberry. Smith 9279. "Nąpųnųpųnųke." A favorite food.

Juglandaceae, Walnut Family

140. *Juglans cinerea* L. Butternut. Smith 9283. "Cazuke" (Fruit tree). (M70, D156)
141. *Juglans nigra* L. Black walnut. [No specimen.] "Cakhu." (M71, D157)

Lamiaceae, Mint Family

142. *Stachys tenuifolia* Willd. (var. *aspera* (Michx.) Fern.) Rough hedge nettle. Smith 9265. "Nįtašjakną'ap" (Indian tea tree leaf). The leaves of this plant are dried and kept for tea, and it substitutes for regular tea from the store.

Liliaceae, Lily Family

143. *Polygonatum biflorum* (Walt.) Ell. (*P. commutatum* (J.A. & J.H. Schultes) A. Dietr.) Solomon's seal. Smith 9342. "šųkhi" (Dog teeth). The root is a vegetable food to the Hocąk. They cook it, then dry and store it for winter and use it in soup.

Oxalidaceae, Wood Sorrel Family

144. *Oxalis stricta* L. Yellow sorrel. Smith 9277. "Xąwįsku" (Weed sweet). To eat. (M81, D158)

Rosaceae, Rose Family

145. *Amelanchier laevis* Wieg. Juneberries. Smith 9178. "Wįxcawux" (Duck's berry). Food.
146. *Crataegus flabellata* (Spach) Kirchn. (*C. grayana* Egglest.) Thorn apple, [Fanleaf-hawthorn]. Smith 9287. "Cosawahu" (Berries on trees). Used as a food.
147. *Fragaria vesca* L. [Thin-leaved wild strawberry.] Smith 9343. "Hastekhu." They eat the fruit.
148. *Prunus serotina* Ehrh. Wild black cherry. Smith 9264. "Nąpakwijanįk" (Tree cherry drunk). This is called the drunk cherry, because if you eat too many you get drunk. (M99)
149. *Prunus virginiana* L. Choke cherry. Smith 9290. "Nąpak" (Tree choke cherry). Food.
150. *Rubus allegheniensis* Porter. Blackberry. Smith 9173. "Hassep" (Black berry). The Indians gather large quantities for food. (M101)
151. *Rubus hispidus* L. [Swamp dewberry]. Smith 9211. "Hassepxųnųnǫk" (Blackberry—little). For food.
152. *Rubus idaeus* L. Red raspberry. Smith 9338. "Hazešucge" (Red raspberries). They eat them. The leaves and vines are used as a tea substitute beverage.

Solanaceae, Nightshade Family

153. *Physalis virginiana* P. Mill. Ground cherry. Smith 9334. "Hąpokhišjasu" (Owl's eyes). They eat the berries.

Vitaceae, Grape Family

154. *Vitis* sp. Wild grape. [No specimen.] "Hapsįc." [Fruits] are eaten. (M115)

Dye Plants

Vascular Plants

Balsaminaceae, Touch-me-not Family

155. *Impatiens capensis* Meerb. (*I. biflora* Walt.) Spotted touch-me-not. Smith 9275. "Xąwįzi." An old time yellow dye plant to dye woven woolen bags or basswood bags.

Juglandaceae, Walnut Family

156. *Juglans cinerea* L. Butternut. Smith 9283. "Cazuke" (Fruit tree). Butternut bark dye is about the same [as walnut], "Cak." The bark is a dye to get a chocolate brown color. (M70, F140)
157. *Juglans nigra* L. Black walnut. [No specimen.] Black walnut dye is "Cak" while the tree itself is "Cakhu." (M71, F141)

Oxalidaceae, Wood Sorrel Family

158. *Oxalis stricta* L. Yellow sorrel. Smith 9277. "Xąwįsku" (Weed sweet). The Sorrel dye is called a tan or buckskin color and the whole plant is boiled to get the color. (M81, F144)

Papaveraceae, Poppy Family

159. *Sanguinaria canadensis* L. Bloodroot. Smith 9367. "Pexhišuc" (Red gourd). This is used for a red or pink dye. (M82)

Polygonaceae, Buckwheat Family

160. *Rumex orbiculatus* A. Gray (*R. britannica* Amer. auct. non L.) [Great water-dock.] [No specimen.] "Hišuc" (Water red). This is another fine red dye and somewhat a tanning agent at the same time. This large root does not occur right here, but the Hocąk can get it at the grist mill at Necedah or at Kilbourn. It is a fine red dye. They say it is their best. (M89)

ALGAE

161. The Hocąk use the blue moss—algae [no specimen] that grows in spring water and sometimes floats to the top. They gather and burn this to make a dye. It is called "maji" (water moss).

FIBER AND MATERIAL USES

VASCULAR PLANTS

Aceraceae, Maple Family

162. *Acer rubrum* L. (Red Maple). Smith 9172. "Wisebu" (Black growing). The wood is used in making large spoons and dishes, especially the dice and bowl game bowls. (F118)

Anacardiaceae, Cashew Family

163. *Rhus hirta* (L.) Sudworth (*R. typhina* L.) Staghorn sumac. Smith 9171. "Hazni̧hu" (Berries, water leaking out). The wood is used for making the flutes with holes in them, love song flutes. The stick is split in half while green and the holes are then cut out in half circles. Then the two halves are glued together. (M2, F120)

Apocynaceae, Dogbane Family

164. Apocynum androsaemifolium L. Spreading dogbane. Smith 9192. "Waruhaxąwi̧" (Fireflies' weed). Late in the fall it furnishes a thread material. (M7)

Asclepiadaceae, Milkweed Family

165. *Asclepias exaltata* L. (*A. phytolaccoides* Pursh). [Tall] Milkweed. Smith 9307. "Mąhi̧c." In the fall this furnishes thread from the rind or bark. (M13, F123)

Betulaceae, Birch Family

166. *Betula papyrifera* Marsh. (*B. alba* L. var. *papyrifera* (Marsh.) Spach.) Canoe birch [Paper birch]. Smith 9185. "Nąhaska" (Tree-white skin). Baskets and canoes are made from the bark. (F126)
167. *Corylus americana* Walt. Hazelnut. Smith 9256. "Huksik" (Indian nut). Used for bag fiber, rough made. (F127)

Cornaceae, Dogwood Family

168. *Cornus racemosa* Lam. (*C. paniculata* L'Hér.) Panicled dogwood. Smith 9263. "Mąsigųsge" (Arrow wood). Its chief use is to make arrows. (M49)

Cyperaceae, Sedge Family

169. *Scirpus cyperinus* (L.) Kunth. Rushes. Smith 9259. "Caiksap." Before the whitemen came, there were no locks, so if the Indians wanted to hide something, they dug a hole, lined it with this rush and concealed it again.

Fabaceae, Bean Family

170. *Gymnocladus dioicus* (L.) K. Koch. Kentucky coffee tree. [No specimen.] "Nąpacaknąkhu." The seeds are used for chips and dice in Indian dice and bowl game. (M61)

Juncaceae, Rush Family

171. *Juncus canadensis* J. Gay *ex* Laharpe. (*J. scirpoides* Lam.) Canadian rush. Smith 9233. "Sa." For making mats.

Oleaceae, Olive Family

172. *Fraxinus nigra* Marsh. Black ash. Smith 9289. "Rakhap." For basket splits.

Osmundaceae, Royal Fern Family

173. *Osmunda claytoniana* L. Ostrich fern [Interrupted fern]. Smith 9179. "Cųkeresge" or "Cųšereke" (Stand up—lots of leaves). Hocąk use it for making a bed when they go hunting. (M80)

Pinaceae, Pine Family

174. *Larix laricina* (DuRoi) K. Koch. Tamarack. Smith 9325. "Casak." Bark wigwams are made from the big ones—1 ½ feet in diameter. You can't find them now. (M83)
175. *Pinus banksiana* Lamb. Jack pine. Smith 9299. "Wazigųsge" (Pine wood). The central part of the root is used to sew birch bark canoes.
176. *Pinus strobus* L. White pine. Smith 9167. "Wazi." The Hocąk preferred the white pines for making big dugout canoes and say that they will last a long time. (M85)

Polygonaceae, Buckwheat Family

177. *Polygonum convolvulus* L. [Black bindweed]. Smith 9248. "Warucžugu" (Growing in field). Used as a string for tying. [Native of Europe.]

Salicaceae, Willow Family

178. *Populus grandidentata* Michx. Large tooth aspen. Smith 9174. "Wasge." Indian children make toys out of it. (M102, A194)
179. *Populus tremuloides* Michx. Trembling aspen. Smith 9175. "Wasge." The use is the same as *P. grandidentata.* (M103, A195)

Tiliaceae, Linden Family

180. *Tilia americana* L. Basswood. Smith 9315. "Hi̧šge." Used to make string, bags, fancy and hunting bags. The wood is used for cradles or baby boards. (M111)

Typhaceae, Cat-tail Family

181. *Typha latifolia* L. Cattails. Smith 9242. "Wicihu" (Leaves mats) or "Kšohi̧" (Baby's coat). Used as a padding or to make mats.

MOSSES

Polytrichaceae

182. *Polytrichum commune* Hedw. (*P. juniperinum* (Willd.) Hedwig.) Smith 9181. "Xa̧je" (Soft, spongy). Used to keep tree [roots] wet a long time before planting.

Sphagnaceae

183. *Sphagnum fallax* (Klinggr.) Klinggr. (*S. warnstorfii* Russ. var. *virescens* Russ. in Warnst.) [Sphagnum moss.] Smith 9182. "Xa̧je" (Soft, spongy). The Indians say it is soft or spongy and use it to keep tree roots wet.

ANIMALS

Glue was made from deerhorn [*Odocoileus* sp., Cervidae], but it doesn't hold long. The best glue is made out of the egg of the sturgeon [*Acipenser fulvescens* Raf., Acipenseridae]. Glue can also be obtained from the back bones of the sturgeon.

ANIMAL FOODS AND MEDICINES

Asteraceae, Sunflower Family

184. *Liatris scariosa* (L.) Willd. Blazing star; Dotted button snakeroot. Smith 9300, 9302. "Cesi̧chošok" (Buffalo's tail). Fed to a weak pony, it fleshes him up. When horses are ill, they feed the root to pep them up. (M28)

Equisetaceae, Horsetail Family

185. *Equisetum arvense* L. [Common horsetail]. Smith 9347. "Hini̧k." Should horses eat too much of this, they would die right away if one ran them. Let them stand for two days. If they eat only a little, they will get fat. The root is horse medicine. The tea acts as a urinary for horses.

Fabaceae, Bean Family

186. *Trifolium pratense* L. Red clover. Smith 9189. "Xa̧wi̧jop" (Four leaved). No use except to feed horses. [Native of Europe.]

187. *Vicia americana* Muhl. *ex* Willd. [American Vetch]. (*Lathyrus* sp.) Smith 9268. "Pišgehuni̧k" (Night hawk bean). Eaten by horses. No [other] use ascribed.

Fagaceae, Beech Family

188. *Quercus rubra* L. Red oak. Smith 9170. "Huksigu" or "Piksigu" (Nut tree). It is used to cure worms in horses. The tincture or tea of this combination is used to soak corn which is fed to the horses to expel worms. (M67, F138)

Liliaceae, Lily Family

189. *Clintonia borealis* (Ait.) Raf. Northern clintonia. Smith 9260. "Ŝųmąką" (Dog medicine). The root is ground up for dog medicine, [in case of] distemper. It is also used to kill worms in a dog. (M75)

Plantaginaceae, Plantain Family

190. *Plantago media* L. Plantain. Smith 9323. Feed a little at a time to horses to cure them. [Native of Europe.] (M87)

Polygalaceae, Milkwort Family

191. *Polygala sanguinea* L. Milkwort. Smith 9226, 9227. "Cesi̧cošok" (Buffalo tail). If horses have distemper, this is cut up and put in their feed. (M88)

Rhamnaceae, Buckthorn Family

192. *Ceanothus americanus* L. [New Jersey tea.] Smith 9332. "Cawaruc" (Deer food). The deer eat the leaves. (M94)

Rubiaceae, Madder Family

193. *Mitchella repens* L. Partridge berry. Smith 9319. [They] don't use the partridge berry and only know that the quail eat the berries.

Salicaceae, Willow Family

194. *Populus grandidentata* Michx. Large tooth aspen. Smith 9174. "Wasge." This bark mixed with red oak was used to cure worms in horses. (M102, C179)

195. *Populus tremuloides* Michx. Trembling aspen. Smith 9175. "Wasge." The use is the same as *P. grandidentata.* (M103, C180)

Smilacaceae, Catbriar Family

196. *Smilax tamnoides* L. (*S. hispida* Muhl.) Bristly sarsaparilla. Smith 9292. "Waxacsep" (Black stickers). When horses are weak, a little of the root at a time, fed once a day in their feed, will build them up. (M110)

NOT USED, OR USE UNKNOWN

VASCULAR PLANTS

Aceraceae, Maple Family

197. *Acer negundo* L. Box elder. Smith 9269. "Na̧hošge" (Tree grows fast). No use for it is known. They merely note the rapidity with which the tree grows in its name.

Asteraceae, Sunflower Family

198. *Arctium minus* Bernh. [Common burdock.] Smith 9274. "Waraxgaxgapke" (stickers). Not used. [Native of Eurasia.]
199. *Artemisia campestris* L. ssp. *caudata* (Michx.) Hall & Clem. (*Artemisia caudata* Michx.) [Wormwood; Sage] Smith 9353. [Consultants] don't know [it].
200. *Bidens coronata* (L.) Britt. (*B. trichosperma* (Michx.) Britt). [Northern tickseed-sunflower.] Smith 9240, 9266. "Rakesep" (Black weed). Don't know.
201. *Conyza canadensis* (L.) Cronq. (*Erigeron canadensis* L.) Horseweed. Smith 9296. "Rake" (Weed). Not used.
202. *Coreopsis* sp. (*C. grandiflora* Hogg. *ex* Sweet). [Tickseed.] Smith 9232. [Specimen is in poor condition and may be *C. verticillata* L. or *C. tinctoria* Nutt.] Don't know.
203. *Coreopsis palmata* Nutt. [Finger-tickseed.] Smith 9330. Don't know it.
204. *Doellingeria umbellata* (P. Mill.) Nees (*Aster umbellatus* P. Mill.) [Tall flat-topped white aster.] Smith 9239. "Poaxų." [No note of use, but species called "Poaxų" are generally used as smudges.]
205. *Euthamia graminifolia* (L.) Nutt. (*Solidago graminifolia* (L.) Salisb.) Fragrant Goldenrod. Smith 9205. "Jiraisagu." No use ascribed to it, unless maybe sweatbath.
206. *Hieracium scabrum* Michx. [Sticky hawkweed.] Smith 9215, 9216. Not known, although some think it is used by others.
207. *Tanacetum vulgare* L. [Tansy.] Smith 9310. [The Hocąk] don't use it. [They] know the whites use it. [Native of the Old World.]

Betulaceae, Birch Family

208. *Betula pumila* L. var. *glandulifera* Regel (*B. glandulifera* (Regel) Butler). Dwarf birch. Smith 9243. "Nąšip" (Tree dwarf). No use.

Campanulaceae, Bellflower Family

209. *Campanula rotundifolia* L. [Harebell.] Smith 9350. [Consultants] don't use it, but some others use the root.

Caprifoliaceae, Honeysuckle Family

210. *Viburnum acerifolium* L. Maple leaved viburnum. Smith 9186. "Nąhųc" (Tree bear). Not used so far as known.

Caryophyllaceae, Pink Family

211. *Silene noctiflora* L. Bladder campion. Smith 9193. "Xąwįhokirukisa" (Close at night). No use discovered. [Native of Europe.]

Commelinaceae, Spiderwort Family

212. *Tradescantia ohiensis* Raf. (*T. reflexa* Raf.) [Smooth spiderwort.] Smith 9337. Don't know.

Cornaceae, Dogwood Family

213. *Cornus canadensis* L. Bunchberry. Smith 9210. "Wakere" (Bad taste). Some use it, but [informant does] not know how.

Cyperaceae, Sedge Family

214. *Eriophorum viridicarinatum* (Engelm.) Fern. [Dark-scale cotton-grass.] Smith 9322. Don't use it.

Equisetaceae, Horsetail Family

215. *Equisetum hyemale* L. [Common scouring-rush.] Smith 9348. [Collected after *E. arvense*, which is a horse medicine; no note about use.]

Ericaceae, Heath Family

216. *Andromeda polifolia* L. var. *glaucophylla* (Link) DC. (*A. glaucophylla* Link). [Bog-rosemary.] Smith 9249. "Nąksikrupak" (Sticks bunch). No use.
217. *Epigaea repens* L. Trailing arbutus. Smith 9326. "Xąwįskapąnąpį" (White smells good). Don't use it.
218. *Gaultheria hispidula* (L.) Muhl. *ex* Bigelow (*Chiogenes hispidula* (L.) Torr.& Gray). [Creeping Snowberry.] Smith 9327. Don't know it.

Fumariaceae, Fumitory Family

219. *Corydalis sempervirens* (L.) Pers. [Corydalis.] Smith 9305. Used but name forgotten.

Geraniaceae, Geranium Family

220. *Geranium bicknellii* Britton. [Wild geranium.] Smith 9298. "Xąwįwašara" (Pretty flowers). Not used.

Lamiaceae, Mint Family

221. *Hedeoma hispidum* Pursh. [Rough false pennyroyal.] Smith 9366. Don't know.
222. *Lycopus uniflorus* Michx. [Northern water-horehound.] Smith 9324. Don't know.

Linaceae, Flax Family

223. *Linum usitatissimum* L. Flax. Smith 9333. Don't use it. [Escaped European cultigen.]

Lycopodiaceae, Clubmoss Family

224. *Lycopodium obscurum* L. [Princess-pine]. Smith 9321. "Wazinįnįsge" (Looks like pine). The Hocąk don't use it, but gather it for white people to use in decorations.

Malvaceae, Mallow Family

225. *Malva rotundifolia* L. [Dwarf mallow.] Smith 9247. "Xąwįwaiskap" (weed bead [?]). No use. [Native of Europe.]

Onagraceae, Evening Primrose Family

226. *Epilobium angustifolium* L. Fire weed. Smith 9199. Said to be a recent migrant to [Hocąk] country

and not known to them. It has been here only five or six years.
227. *Epilobium coloratum* Biehler. [Eastern willow-herb.] Smith 9237. Don't know.
228. *Oenothera biennis* L. Evening primrose. Smith 9194. "Raxgesuserec" (Long seed). No use discovered.

Orchidaceae, Orchid Family

229. *Malaxis unifolia* Michx. (*Microstylis unifolia* (Michx.) BSP.) [Green adder's mouth.] Smith 9213. No use.
230. *Spiranthes cernua* (L.) Richart. [Nodding ladies-tresses.] Smith 9229. Don't know.

Poaceae, Grass Family

231. *Hordeum jubatum* L. Squirrel tail grass. Smith 9254. "Xąwįsįcserec" (Weed long tail). While they have a name for this, it has no use in any manner.

Polygalaceae, Milkwort Family

232. *Polygala polygama* Walt. [Bitter milkwort]. Smith 9344. They don't know it.

Polygonaceae, Buckwheat Family

233. *Polygonum amphibium* L. (*Polygonum muehlenbergii* (Meisn.) Wats.) [Water smartweed.] Smith 9245. "Xąwįokerešuc" (Weed red flowers). No use.
234. *Polygonum arifolium* L. [Halberd-leaved tear-thumb.] Smith 9246. "Xąwįmąįsu" (Yellow head). No use.
235. *Polygonum pensylvanicum* L. (var. *laevigatum* Fernald, f. *pallescens* Stanford) [Pennsylvania smart-weed.] Smith 9228. Don't use it.
236. *Polygonum sagittatum* L. [Arrow-leaved tear-thumb.] Smith 9219. "Waraxgapxgapke" (Sticky). Not used.

Primulaceae, Primrose Family

237. *Lysimachia quadriflora* Sims. (*Steironema quadriflorum* (Sims.) Hitchcock). Fourleaved loosestrife. Smith 9197. "Raxge'apjop" (Four leaves). [There is] a use, but not known to informant.
238. *Trientalis borealis* Raf. (*Trientalis americana* Pursh). [Starflower.] Smith 9329. Don't know.

Pyrolaceae, Shinleaf Family

239. *Chimaphila umbellata* (L.) Barton. [Prince's pine.] Smith 9346. Don't know it.

Ranunculaceae, Buttercup Family

240. *Clematis virginiana* L. [Virgin's bower.] Smith 9311. Don't know it.

Rosaceae, Rose Family

241. *Agrimonia gryposepala* Wallr. [Common agrimony.] Smith 9295. "Waraxgaxgapke" (Stickers). No use.
242. *Spiraea alba* DuRoi (*S. salicifolia* L.) [Meadowsweet.] Smith 9230. Don't know.
243. *Spiraea tomentosa* L. var. *rosea* (Raf.) Fern. (*S. tomentosa* L.) [Hardhack.] Smith 9231. Don't know.

Rubiaceae, Madder Family

244. *Galium concinnum* Torr. & Gray. [Shining bedstraw.] Smith 9352. Don't know.

Rutaceae, Rue Family

245. *Zanthoxylum americanum* P. Mill. Prickly ash. [No specimen.] "Nąpahi" (Tree stick sharp). They use it, too, but none grows right close to Monegar's.

Scrophulariaceae, Figwort Family

246. *Linaria vulgaris* Miller. [Butter-and-eggs.] Smith 9349. Don't know. [Native of Europe.]

Vitaceae, Grape Family

247. *Parthenocissus quinquefolia* (L.) Planch. (*Psedera quinquefolia* (L.) Greene). Virginia creeper. Smith 9286. "Ną'apxete" (Big leaves). Not used.

Acknowledgements

This project has been greatly assisted by many people. At the Milwaukee Public Museum, Neil Luebke, Curator of Vascular Plants, provided invaluable assistance with the manuscript and plant species identification, and interns John Christy and Kevin Lyman helped with verification of herbarium specimens. Ann McMullen, Curator of North American Ethnology, helped us find appropriate photographic and other archived material and reviewed the manuscript. Nancy Oestreich Lurie commented extensively on the manuscript and provided invaluable insight into Hocąk culture.

We are grateful to the Hocąk Nation for their participation in this project. Anna Wilson, member of the Hocąk Nation and University of Kansas biology student, served as our liaison with the Hocąk Nation. Her assistance in contacting tribal entities and providing other information and support made publication of this work possible. At the Hocąk Wazijaci Language and Culture Program, Sheila Shigley and Kenneth Funmaker, Sr. corrected the spelling of Hocąk plant names and informed us of the history of the Hocąk people and the meaning of their name. Kenneth Funmaker also provided the Hocąk translation of his statement. Nettie Kingsley, Director of Historic Preservation, assisted with historical information.

At the University of Kansas, Robert Rankin informed us of the existence of Smith's Hocąk manuscript and gave us our first photocopy. Alexandra Fraser typed the handwritten field notes, and Jill Hanlon, Lisa Kahn and Jeremy McLain edited the manuscript. Mary Adair, Ray Pierotti, and Paul Rich reviewed and edited the manuscript. In addition, the manuscript benefitted from outside review from the following colleagues: M. Kat Anderson, National Plant Data Center, Natural Resources Conservation Service, Department of Environmental Studies, University of California at Santa Cruz; Chuck Haines, Haskell Indian Nations University; and Nancy Turner, Environmental Studies Program, University of Victoria. The Madison and Lila Self Graduate Fellowship Program provided Dana Hurlburt's support while she worked on this project.

Original assistance for the project was provided by Huron Smith's colleagues at the Milwaukee Public Museum, including: William C. McKern, Associate Curator of Anthropology; Towne L. Miller, Honorary Curator of Archaeology; Samuel A. Barrett, Director.

This work depended on the generosity and good will of the Hocąk people with whom Smith originally worked, including: John Bear, the family of Dave Dekorah, Henry Dick, "Cahabįge" (Annabelle Eagle), Mr. and Mrs. Jim Henniger, Mr. and Mrs. Fred Mallory, "Iwahonąžįka" and Beaver Woman (Mr. and Mrs. George Monegar), Mr. and Mrs. Frank Smith, Ray White, and Mr. and Mrs. Ulysses A. White. Ultimately, we need to acknowledge the Hocąk Nation for bringing this knowledge to all of us.

Literature Cited

Abbott, L. R. 1988. Frontier lead mining in the upper Mississippi valley. Journal of the Iowa Archaeological Society 35:1–15.

Anderson, L. E. 1990. A checklist of Sphagnum in North America north of Mexico. The Bryologist 93:500–501.

Anderson, L. E., H. A. Crum, and W. R. Buck. 1990. List of the mosses of North America north of Mexico. The Bryologist 93:448–499.

Andros, F. 1883. The medicine and surgery of the Winnebago and Dakota Indians. American Medical Association Journal 1:116–118.

Barrett, S. A. 1933a. In Memoriam. Bulletin of the Public Museum of the City of Milwaukee 7:4.

Barrett, S. A. 1933b. Huron Herbert Smith 1883–1933. Wisconsin Archeologist 12:69–74.

Bigony, B. A. 1982. Folk literature as an ethnohistorical device: The interrelationships between Winnebago folk tales and Wisconsin habitat. Ethnohistory 29:155–180.

Curtis, J. T. 1959. The vegetation of Wisconsin. University of Wisconsin press, Madison.

De La Ronde, J. T. 1876. Personal narrative. Collections of the State Historical Society of Wisconsin 7:345–365.

Dorsey, J. O. and P. Radin. 1910. Winnebago. Pages 958–961 *in* F. W. Hodge, ed. Handbook of American Indians north of Mexico. Bureau of American Ethnology Bulletin 30, part 2. Smithsonian Institution, Washington, D.C.

Fay, G. E. 1966. Treaties between the Winnebago Indians and the United States of America. Journal of the Wisconsin Indians Research Institute 2(1):7–49.

Funmaker, A. R. 1983. Effects of dietary and ecological changes on Native American health: the case of the Winnebago. California Anthropologist 13: 15–20.

Gilmore, M. R. 1991. Uses of plants by the Indians of the Missouri River region. University of Nebraska Press, Lincoln. Reprinted from the Thirty-third Annual Report of the Bureau of American Ethnology, published by the Government Printing Office, Washington, D.C., in 1919.

Gleason, H. A. and A. Cronquist. 1991. Manual of vascular plants of Northeastern United States and adjacent Canada. Second ed. New York Botanical Garden, Bronx.

Gray, A. 1908. Gray's New Manual of Botany, a handbook of the flowering plants and ferns of the central and northeastern United States and adjacent Canada. 7th ed., rearranged and revised by B. L. Robinson and M. L. Fernald. American Book Company, New York.

Hill, T. W. 1990. Peyotism and the control of heavy drinking: the Nebraska Winnebago in the early 1900s. Human Organization 49:255–265.

Hocąk Nation Department of Historical Preservation. 1996. The HoChunk Nation.

Jenks, A. E. 1900. Wild rice gatherers of the Upper Lakes. Pages 1013–1137 *in* Bureau of American Ethnology 19th Annual Report. Government Printing Office, Washington, D.C.

Kartesz, J. T. 1996. A synonymized checklist of the vascular flora of the United States, Canada, and Greenland. Revision. The Biota of North America Program of the North Carolina Botanical Garden, University of North Carolina Press, Chapel Hill.

Kay, J. 1984. The fur trade and Native American population growth. Ethnohistory 31:265–287.

Lawson, P. V. 1907. Habitat of the Winnebago, 1632–1832. Proceedings of the State Historical Society of Wisconsin 54:144–166.

Lurie, N. O. 1960. Winnebago protohistory. Pages 790–808 *in* S. Diamond, ed. Culture in history: Essays in honor of Paul Radin. Columbia University Press, New York.

———. 1961. Mountain Wolf Woman, sister of Crashing Thunder: The autobiography of a Winnebago Indian. University of Michigan Press, Ann Arbor.

———. 1966. A check list of treaty signers by clan affiliation. Journal of the Wisconsin Indians Research Institute 2(1):50–79.

———. 1978. Winnebago. Pages 690–707 *in* B. G. Trigger, ed. Handbook of North American Indians. Vol. 15: Northeast. Smithsonian Institution, Washington, D.C.

Merrell, H. 1876. Pioneer life in Wisconsin. Collections of the State Historical Society of Wisconsin 7:367–404.

Merry, C. A. and W. Green. 1989. Sources for Winnebago history in Northeastern Iowa, 1837–1848. Journal of the Iowa Archaeological Society 36:1–8.

Peske, G. R. 1971. Winnebago cultural adaptation to the Fox River waterway. Wisconsin Archaeologist 52(2):62–70.

Radin, P. 1990. The Winnebago tribe. University of Nebraska Press, Lincoln. Reprinted from the Thirty-seventh Annual Report of the Bureau of American Ethnology, published by the Smithsonian Institution, Washington, D.C., in 1923.

Schoolcraft, H. R. 1975. Personal memoirs of a residence of thirty years with the Indian tribes on the American frontiers. Arno Press, New York. Reprint of 1851 edition published by Lippincott, Grambo and Co., Philadelphia.

Smith, A. E. and V. Carstensen. 1974. Economic and historical background for the Winnebago Indian claims. Pages 225–456 *in* D. A. Horr, ed. Winnebago Indians. Garland Series in Ethnohistory, North Central and Northeastern Indians. Garland Publishing, New York.

Smith, A. H., H. V. Smith, and N. S. Weber. 1981.

How to know the non-gilled mushrooms. Wm. C. Brown Co., Dubuque, Iowa.
Smith, Huron H. 1921. Ethno-botanical Collecting on the Menomini Reservation. Yearbook of the Public Museum of the City of Milwaukee 1:50–56.
———. 1923. Ethnobotany of the Menomini Indians. Bulletin of the Public Museum of the City of Milwaukee 4:1–174.
———. 1928a. Ethnobotany of the Meskwaki Indians. Bulletin of the Public Museum of the City of Milwaukee 4:175–326.
———. 1928b. Among the Winnebago. Yearbook of the Public Museum of the City of Milwaukee 8: 76–82.
———. 1932. Ethnobotany of the Ojibwe Indians. Bulletin of the Public Museum of the City of Milwaukee 4:327–525.
———. 1933. Ethnobotany of the Forest Potawatomi Indians. Bulletin of the Public Museum of the City of Milwaukee 7:6–230.
Thwaites, R. G., ed. 1892. The Wisconsin Winnebagoes. An interview with Moses Paquette. Collections of the State Historical Society of Wisconsin 12:399–433.
———. 1902. The French regime in Wisconsin-I. Collections of the State Historical Society of Wisconsin 16:2–477.

Book Review

At the Desert's Green Edge. An Ethnobotany of the Gila River Pima. Amadeo M. Rea. 1997. The University of Arizona Press, 1230 Park Ave., Tucson, AZ 85719. 430 pp. (hardcover). $60.00. ISBN 0-8165-1540-9.

This book began in 1963 when the author went to the Arizona Pima reservation to teach high school. Rea, an ornithologist by training, wrote his first book *Once a River* (1983, University Arizona Press, Tucson) detailing recent habitat changes on the reservation and consequent impacts on birds and plants. During those years he became better acquainted with the people and their plants. *At the Desert's Green Edge* is the phenomenal result.

Published in a coffee-table size (9×12 in.), three-column text, with black and white photographs, line drawings of habitats and plant parts, and illustrated by Takashi Ijichi's *sumi-e* paintings, this is a volume to behold. Yet, those are just the surface impressions. Within is "The Pima and Their Country" (pp. 5–90), "Gila Pima Plants" (pp. 91–376), four appendices (pp. 377–400), a bibliography (pp. 401–411), and index. Rea chose to continue calling the people by their "outside" name. Pima is the name usually associated with them, and simplifies a complex array of group relationships that he discusses in detail (pp. 5–14).

Part one describes the people, the changes in their habitat including the draining of their river, and basic information about their language. Next, Rea discusses the Pima classification of plants, and lists them by their *O'odham* (the people's word for themselves) names. In doing this, he provides us with a sketch of their worldview. Some of their plant groups are *Vashai* (grasses & grassy plants), *Livagi* (eaten greens), and *E'es* (crops). Within the groups, the species are arranged alphabetically by genus. All organisms are listed by common and scientific name in the index. This makes the book useful to *O'odham* and non-*O'odham*.

The book includes 250 plants that are (or recently were) used by these people. Each entry puts the species in historical context. For example, *kuup chevel-dakud* is "hairlike algae growing in rivers." Although this alga was once common in Pima country, its name is not recognized by some now. On the devastatingly arid land of the modern Pima reservation, it is hard to believe that these were once a riverine people living in the same places by agriculture and fishing. Much more detail is given for plants like *hui* (*Prosopis velutina*), ki'aki (*Amaranthus hibridus*), *ko'okol* (*Capsicum* spp.), *haal* (*Cucurbita* spp.), and *huun* (*Zea mays*).

Amadeo Rea and I met in 1989. He impressed me as unusually sensitive to the Pima, their lands, and their ways. We also talked about plants and animals, and he was knowledgeable about all of these topics. Those impressions are reinforced on each page of this book. Amadeo has achieved the nearly impossible. He has salvaged the remaining collective knowledge of a fragmented and partly acculturated people and presented it in a remarkably palatable and useful format. This is an outstanding book that adds another key piece of the Sonoran human and habitat puzzle. Every person interested in American people and plants should use this book.

Daniel F. Austin
Department of Biological Sciences
Florida Atlantic University
Boca Raton, FL 33431

CHEMICAL COMPOSITION, NUTRITIONAL EVALUATION, AND ECONOMIC PROSPECTS OF *SPONDIAS PURPUREA* (ANACARDIACEAE)[1]

MICHAEL J. KOZIOŁ AND MANUEL J. MACÍA

Kozioł, Michael J. (*Nestlé R&D Center S.A., Casilla Postal 17-11-6053, Quito, Ecuador*) *and* **Macía, Manuel J.** (*Real Jardín Botánico de Madrid, Plaza de Murillo 2, 28014 Madrid, España. Current address: Herbario QCA, Departamento de Ciencias Biológicas, Pontificia Universidad Católica del Ecuador, Casilla Postal 17-01-2184, Quito, Ecuador*). CHEMICAL COMPOSITION, NUTRITIONAL EVALUATION, AND ECONOMIC PROSPECTS OF *SPONDIAS PURPUREA* (ANACARDIACEAE). *Economic Botany 52(4):373–380. 1998. The fruits of "ovo"* (Spondias purpurea *L.) had the highest caloric density of the fruits compared, 74 kcal/100 g edible portion versus 39 to 58 kcal/100 g for peach, apricot, plum, mango, and cherry. This higher caloric density is due principally to ovo's higher concentration of total carbohydrates (19.1%); fructose, glucose, and sucrose together account for 65% of the soluble matter. Unlike the other fruits, ovo retains a fair amount of starch in the mesocarp. It is a moderate source of potassium (250 mg/100 g edible portion) and an excellent source of vitamin C (49 mg/100 g edible portion). Analysis of volatile flavor compounds showed 2-hexenal to be the main flavor compound present. Although there is a local market for ovo in Ecuador, large scale exportation of the fresh fruit currently lacks the necessary infrastructure to prolong shelf life, while the current production of ovo is insufficient for large scale industrial processing of the fruit. In the near future, the best alternative to maximize the income of the small holders producing ovo is for them to concentrate on producing "value added products" such as jams, ice creams, alcoholic beverages, and vinegars.*

COMPOSICIÓN QUÍMICA, EVALUACIÓN NUTRICIONAL, Y PERSPECTIVAS ECONÓMICAS DE *SPONDIAS PURPUREA* L. (ANACARDIACEAE). *Los frutos del "ovo"* (Spondias purpurea *L.) tienen el valor más alto de densidad calórica entre los frutos comparados, 74 kcal/100 g de porción comestible frente a valores de 39 a 58 kcal/g reportados para durazno (melocotón), albaricoque, ciruela, mango y cereza (guinda). Este valor se debe principalmente a la elevada concentración de los hidratos de carbono (19.1%) en el ovo, en comparación con las otras frutas mencionadas; la concentración total de fructosa, glucosa y sacarosa representa el 65% de los sólidos totales presentes. En relación con las otras frutas, el ovo cuenta con cantidades relativamente altas de almidón en el mesocarpio. El ovo representa una fuente regular de potasio (250 mg/100 g de porción comestible) y una excelente fuente de vitamina C (49 mg/100 g de porción comestible). El análisis de los compuestos aromáticos volátiles identificó a 2-hexenal como el componente principal del aroma de la fruta. En Ecuador existe un mercado potencial para la venta de los ovos, aunque en la actualidad, la exportación a gran escala de los frutos frescos carece de la infraestructura necesaria para prolongar la estabilidad del fruto para su transporte y su venta posterior. La mejor alternativa para obtener mayores ingresos para los pequeños productores de ovo sería la venta de los productos que aportan un valor añadido: mermeladas, helados, bebidas alcohólicas y vinagres. Con la producción actual de ovo en Ecuador, sería posible una explotación a nivel de pequeña industria de estos productos elaborados.*

Key Words: *Spondias purpurea*; Anacardiaceae; edible fruit; composition; nutritional value; economic botany; Ecuador.

Two species of the pantropical genus *Spondias* (Anacardiaceae) are native to Ecuador: *S. mombin* L. which is highly polymorphic (Barfod 1987) and occurs on both sides of the Andes, and *S. purpurea* L. which is a well defined species distributed mainly on the western coastal plain of the Andean ridge. Locally, the fruits of both species of *Spondias* are most commonly and indiscriminately called "ovo." *Spondias*

[1] Received 3 June 1997; accepted 4 April 1998.

mombin, which has just begun to appear for sale in the markets in Quito, is the larger and more yellow of the two and is generally acknowledged as being bittersweet whereas the better known *S. purpurea* is smaller, orange to red colored, sweeter and the more aromatic of the two. Ovos are usually consumed fresh although they can be further processed into jams, ice creams or alcoholic beverages (for general information on the description, cultivation, and economic botany of *S. purpurea* L. see León and Shaw 1990 and Macía 1997). Two "schools" exist regarding the method of consumption of ovo: the fruit with peel is eaten whole later spitting out the rather large seeds, or the fruit pulp is softened by gently squeezing the whole fruit between the fingers followed by biting a hole in the peel at one end of the fruit and sucking out the juices and softened pulp.

The present study was undertaken to complement the information of Macía (1997), as well as to enhance the existing data on the chemical composition of *S. purpurea* and to assess its nutritional quality. In the following text, "ovo" will refer specifically to *S. purpurea.*

Materials and Methods

Ripe fruits (5 kg) were collected during the peak of the harvest at the end of March 1995 in the village of Ambuquí, Province of Imbabura, in the north of Ecuador. This village has two areas of production, "sector del pueblo" and "sector de la playa." In "sector del pueblo" a 1150 m^2 plot was selected containing 117 trees which produced 750 kg of ovo in 1995; only 25% (188 kg) of this production was harvested as tree-ripened fruit, the remainder having been harvested green for transport to market (Macía 1997). As our sample represented 2.7% of the production of ripe ovo produced in "sector del pueblo" plot we consider it a valid sampling of this one area, but not of ovo produced nationally. Given the lack of current large scale industrial interest in processing the fruit a more rigorous sampling and replication of analyses to determine the variation in the quality of ovo produced nationally could not be justified.

Upon arrival in the laboratory, samples of pulp and peel were taken for immediate analysis of moisture, pH, °Brix and aromatic compounds. Pulp and peel separated from the remaining fruits were combined and freeze-dried, ground to the consistency of flour and stored at −80°C until analysis. All values obtained for freeze-dried material were subsequently corrected to a fresh weight basis. With the exception of analysis of the aromatic volatiles in ovo, all analyses were performed on duplicate samples of plant material.

Moisture content was determined for fresh and freeze-dried plant material by oven drying for 4 h at 102°C, fat by Mojonnier-type extraction with diethyl ether, protein as Kjeldahl nitrogen × 6.25 (Kjeldahl nitrogen determined with Büchi digestion and distillation apparatus, Büchi Laboratoriums-Technik AG, Flawil, Switzerland), and minerals (ash) by incineration at 550°C for 4 h. Crude fiber was determined on defatted samples using the Fibertec system (Tecator A. B., Höganäs, Sweden) which involved a sequential extraction with 1.25% H_2SO_4 then 1.25% NaOH, drying the residue for 4 h at 102°C followed by incineration for 30 min at 600°C; the difference between the dry weight and ash content of the residue is taken as an estimation of the crude fiber content. Total carbohydrates were calculated as the difference, i.e., 100% − (% moisture) − (% protein) − (% fat) − (% ash). The caloric value per 100 g edible portion was calculated according to the system of Atwater, namely: kcal = (3.36 × % protein) + (3.60 × % total carbohydrate) + (8.37 × % fat) (Merrill and Watt 1973). The pH of the pulp was measured directly and the °Brix determined at 20°C with an Atago refractometer.

Starch was determined enzymatically according to the procedure described by Boehringer Mannheim (1989). Plant material for pectin analysis was extracted three times with boiling 80% ethanol to remove compounds which could interfere with the analysis. Residual ethanol, which can also interfere with the analysis, was removed azeotropically with diethyl ether. Total pectic substances were then extracted three times with 0.5% EDTA and quantified according to the procedure of McCready and McComb (1952).

Fructose, glucose, sucrose, and organic acids were determined by HPLC. The lyophilized plant sample was extracted for 20 min with distilled and deionized water at 60°C, neutralized to pH 7 and brought to 100 ml. After centrifugation, 20 ml of the supernatant liquid was passed through a column containing 4 ml of Dowex 1-X8 anionic exchange resin (100–200 mesh), and the resin rinsed with 5 ml of distilled and deionized water. The sample plus the rinse were then passed through a column containing 4 ml of Dowex 50W-X8 cationic exchange resin (100–200 mesh). The resin was rinsed with 5 ml of distilled and deionized water, and the rinse added to the eluate. The eluate was treated with Carrez solutions to precipitate any proteins present, adjusting the final volume to 50 ml. After centrifugation, an aliquot of the supernatant liquid was filtered (micropore filter, 0.2 μm) and 20 μl injected onto a Bio-Sil Amino 5S column (Bio-Rad) for the analysis of sugars. The mobile phase was acetonitrile: water (70:30) at a flow of 1.0 ml min^{-1} and 6.6 MPa: sugars were detected using a refractive index detector (Perkin-Elmer LC-25 RI detector; range dRI $\times 10^{-6} = 50$, at 20°C).

Organic acids were eluted from the Dowex 1-X8 column with 10 ml of 2 N HCl. An aliquot of this eluate was filtered (micropore filter, 0.2 μm) and 20 μl injected onto an Aminex Ion Exclusion HPX-87H column (Bio-Rad) for the analysis of organic acids. The mobile phase was 0.01 N H_2SO_4 with 0.01% sodium azide at a flow rate of 0.6 ml min^{-1} and 4.2 MPa: organic acids were detected using a refractive index detector (Perkin-Elmer LC-25 RI detector; range dRI $\times 10^{-6} = 50$, at 20°C).

Minerals were determined by flame atomic-absorption spectrophotometry (Model 4000, Perkin-Elmer, Norwalk, CT, USA), adding a solution of lanthanum chloride for the analysis of calcium.

Aromatic volatiles were determined by simultaneous steam distillation-extraction of 560 g of ovo pulp + skin for 90 min using water and *n*-pentane (method of Nickerson and Likens 1966, as modified by Maarse and Kepner 1970). For analysis, 5 μl of the organic fraction were injected into a Hewlett Packard 5890 Series II gas chromatograph coupled to a 5970 mass selective detector. Instrumental conditions were: injector temperature = 200°C; helium linear flow = 28.2 cm s^{-1}; and column = Hewlett-Packard Ultra 2, 25 m × 0.2 mm diameter with a film thickness of 0.33 μm; impact energy = 70 eV; tuning was by autotune using perfluorotributylamine. The temperature program was: hold 50°C for 1 min, increase temperature by 3°C min^{-1} to 160°C, hold 160°C for 2 min, increase temperature by 2°C min^{-1} to 220°C then by 10°C min^{-1} to 290°C, hold 290°C for 2 min. Compounds were tentatively identified by comparison of their mass spectra to those of pure compounds stored in a user-created library. Results for individual compounds are reported as their percentage of the total area of peaks in the total ion chromatogram ("abundance" in Table 2).

RESULTS AND DISCUSSION

The values for the following analyses performed on ovo in this study are included within the ranges given in Table 1: moisture 79.7%, protein 0.9%, fat 0.24%, fiber 0.4%, ash 0.7%, total carbohydrates 18.1%, food energy 70 kcal, sucrose 7.21%, calcium 17 mg, iron 0.30 mg, sodium 9 mg, and potassium 230 mg. In addition to an analysis of the aromatic volatiles in ovo, this study also contributes values for starch, pectins, fructose, glucose, zinc, pH, °Brix, citric acid, malic acid, oxalic acid, and tartaric acid, not previously reported for ovo.

The ripe fruits of ovo most closely resemble over-ripe plums, and for this reason the nutritional evaluation was performed as a comparison with other drupe fruits, four from the family Rosaceae (plum, apricot, cherry, and peach) and one from Anacardiaceae (mango; Table 1). In ovo as purchased, the seed accounts for 34% of the weight of the fruit and the peel another 8% (Leung and Flores 1961): the pulp can account from 50–58% of fresh fruit weight (Leung and Flores 1961, Winton and Winton 1935).

In describing the food uses of *S. purpurea* Morton (1987) stated "while [the ripe fruits are] not of high quality, they are popular with people who have enjoyed them from childhood," yet ovo shows the highest energy value of the fruits compared in this study, due principally to its higher concentrations of total carbohydrates (Table 1). The fruit is rather sweet as attested by its high sucrose and fructose concentrations. The total concentrations of the three sugars quantified account for 65% of the total soluble solids measured as °Brix. The fiber content of ovo is uncharacteristically low for fruits, while there is a considerable amount of starch in the ripe fruit (the starch content of the unripe fruit is higher still, at 8.45 g per 100 g edible portion). The free sugars and starch are readily available as fermentable substrates, which is taken advantage of by local people to produce a very refreshing and incredibly effervescent wine.

Although the pectin content is rather low, it is still sufficient for producing a jam without the need to add more pectin. Given the acidity of

TABLE 1. COMPARISON OF CHEMICAL COMPOSITION OF OVO (*S. PURPUREA*) WITH OTHER FRUITS, PER 100 G EDIBLE PORTION.

Component	Unit	Ovo Range	Ovo Average	Ovo Ref[b]	Other drupe fruits[a] Plum	Apricot	Cherry	Peach	Mango
Moisture	g	65.0–87.0	77.6 (9)	[1–8]	83.7	85.3	82.8	87.5	82.0
Protein	g	0.1–1.0	0.7 (9)	[1–8]	0.6	0.9	0.9	0.8	0.6
Fat	g	0.03–0.8	0.2 (7)	[1–4, 6]	0.2	0.1	0.4	0.1	0.3
Fiber	g	0.2–0.7	0.5 (8)	[1–7]	1.7	1.9*	1.9	0.8	1.7
Minerals (ash)	g	0.3–1.1	0.7 (9)	[1–8]	0.5	0.7	0.5	0.5	0.5
Total carbohydrates	g	16.0–22.3	19.1 (5)	[1–5]	11.9	9.9	12.7	8.7	12.8
Food energy	kcal	61–86	74 (5)	[1–5]	52	44	58	39	56
Starch	g		2.47	[1]	0*	traces	0*	0*	0.3*
Pectins	g		0.22	[1]	0.8	1.0	0.4	0.5	
Fructose	g		2.53	[1]	2.1	0.9	5.5	1.3	2.6
Glucose	g		2.00	[1]	2.7	1.7	6.1	1.2	0.9
Reducing sugars	g		8.08	[8]					
Sucrose	g	5.97–7.21	6.59 (2)	[1,8]	2.8	5.1	0.2	5.4	9.0
Calcium	mg	6–25	17 (8)	[1–7]	14	16	17	8	12
Phosphorus	mg	32–56	42 (7)	[2–7]	18	21	20	23	13
Iron	mg	0.09–1.22	0.72 (8)	[1–7]	0.4	0.7	0.4	0.5	0.4
Sodium	mg	2–9	6 (2)	[1,3]	2	2	3	1	5
Potassium	mg	230–270	250 (2)	[1,3]	221	278	229	205	190
Zinc	μg		20	[1]	70	70	150	20	100*
Carotene[c]	μg	4–225	119 (6)	[2–7]	210	1790	84	440	2770
Thiamine	μg	33–110	84 (7)	[2–7]	70	40	40	30	50
Riboflavin	μg	14–80	40 (7)	[2–7]	40	50	40	50	50
Niacin	mg	0.4–1.8	1.0 (7)	[2–7]	0.4	0.8	0.3	0.9	0.7
Ascorbic acid	mg	26–73	49 (7)	[2–7]	5	9	15	10	39
pH			3.29	[1]					
°Brix			18	[1]					
Citric acid	mg		30	[1]	34	400	13	240	296
Malic acid	mg		110	[1]	1220	1000	940	330	74
Oxalic acid	mg		30	[1]	12	3	7	0	36
Tartaric acid	mg		20	[1]			traces		81

[a] Data from Souci, Fachmann, and Kraut (1986) for plum (*Prunus domestica*), apricot (*Prunus armeniaca*), cherry (*Prunus avium*), peach (*Prunus persica*), and mango (*Magnifera indica*), respectively, except data marked with an asterisk (*) which are from McCance and Widdowson (1991).

[b] Source of data for ovo: 1, this study: 2, División de Investigaciones en Alimentos (1983) for "ciruela de huesito"; 3, Duke and Atchley (1986) for *Spondias purpurea*, citing data from Leung, Butrum and Chang (1972); 4, Instituto Nacional de Nutrición (1965) for "ciruela hobo"; 5, Leung and Flores (1961) for "mombin, purple ('jobo')"; 6, Morton (1987) for "purple mombin"; 7, Sturrock (1959) for "red mombin"; and 8, Winton and Winton (1935) for "red mombin". Some of the published sources listed single values, others a range of values. Where ranges were reported, the high and low values were included in the calculation of the "Average" given in the Table, and the number of values averaged given in parenthesis after each value. Where a range of values was averaged, the number of values averaged will be greater than the number of authors cited.

[c] As the lowest value reported by Morton (1987) for carotene (4 μg) is almost 17 times lower than the next lowest value reported in the literature (66 μg both by the División de Investigaciones en Alimentos 1983 and by Sturrock 1959) we must assume some error either in reporting or in analysis. This suspiciously low value has therefore been excluded in the calculation of the average value of carotene. To standardize the presentation of the data, carotene content was calculated from the vitamin A value reported by the Instituto Nacional de Nutrición (1965) and from the retinol equivalents reported by the División de Investigaciones en Alimentos (1983).

the fruit (pH = 3.3) and the formation of jams without additional gelling agents, it is most likely that native pectins are of the "high methoxyl" type, i.e., with degrees of esterification in excess of 50% (Mitchell, Buckley, and Burrows 1978).

Ovo is a moderate source of potassium (i.e., providing between 100 and 300 mg per serving; Guthrie 1979). Although potassium is recognized as a dietary essential, no minimal requirements have been established; rather, requirements for growth per day are estimated to average 15–20 mg/kg body mass for children 1–10 years old and 35 mg/kg body mass for adolescents (National Research Council 1989). We calculate that a 100 g edible portion of ovo would provide 63% of the potassium requirements for children 4–6 years old, 44% for children 7–10 years old, 16% for adolescents 11–14

TABLE 2. VOLATILE FLAVOR COMPOUNDS IDENTIFIED IN *S. PURPUREA*.

Compound	Abundance	Odor quality
Hexanal	6.95	Fatty, green grass-like[a]
2-Hexenal	38.99	Green, fruity, vegetable-like[a]
3-Methylbutyl acetate	0.22	Fruity, pear-like, banana-like[a]
4-Penten-1yl-acetate	1.29	
3-Methyl-2-buten-1-ol acetate	1.30	
Ethyl 2-methyl but-2-enoate	0.19	
2-Heptenal	0.07	Pungent green, somewhat fatty odor, pleasant only on extreme dilution[c]
2,6,6–Trimethyl-2-ethylenetetrahydro-2H-pyran	0.20	
2,4-Heptadienal	0.08	Rotten apple, rancid, hazelnut[d]
Ethyl hexanoate	0.89	Fruity, floral, wine-like[a]; powerful fruity with pineapple-banana note[b]
3-Hexen-1-ol acetate	0.88	
Trans-Ocimene	0.03	Warm, herbaceous[a]
Phenylacetaldehyde	0.04	Pungent green, floral, sweet, wallflowers-like[a]
α-Pinene	0.12	Floral, warm, resinous, pine, cedar-wood like[a]
Linalool oxide	0.40	Sweet, woody, floral, earthy[a]
Methyl benzoate	0.76	Fruity, similar to cananga[c]
α-Terpinolene	1.19	Floral, sweet, pine-like[a]
Methyl 4-pyridinecarboxylate (methyl isonicotinate)	0.05	
Ethyl benzoate	0.21	Blackcurrant-like[a]
Methyl 2-hydroxybenzoate	0.49	
Ethyl octanoate	0.45	Apricot-like, banana-like, wine-like[a]
3-Ethyl-5-methyl-1-propyl-cyclohexane	0.26	
1,1,6-Trimethyl-1,2-dihydronaphthalene	0.11	
Ethyl decanoate	0.47	Floral, sweet, oily nut-like, brandy-like[a]
Ethyl dodecanoate	0.55	Faint fruity, flower petal-like, slightly mango-like[a]
Tetradecanoic acid	3.07	Faint waxy, oily[a]
Ethyl tetradecanoate	0.66	Oily, ethereal, violet-like, orris-like[a]
1,2-Benzenedicarboxylic acid	0.52	
Hexadecanoic acid	18.51	Virtually odorless[a]

Values for "abundance" are the percentages of the area of the total ion chromatagram represented by the peaks of each of the compounds identified. References cited as to the odor characteristics of the compounds identified are:
[a] Shibamoto and Tang (1990).
[b] Engel, Heidlas and Tressl (1990).
[c] Furia and Bellanca (1975).
[d] Kochhar (1996).

years old, 12% for adolescents 15–18 years old, and 10% for adults.

The vitamin C (ascorbic acid) content of ovo is the highest of the fruits compared, and a 100 g edible portion would provide 98–123% of the recommended dietary allowance (RDA) for children 1–14 years old, and 82% of the RDA for people over 14 years old (National Research Council 1989).

Table 2 presents some of the volatile compounds we were able to identify by comparison of their mass spectra with those in our mass spectrum library of volatile aromatics; where the information was available, the literature reports of the odor characteristics of these compounds is also given. The major aromatic compound was 2-hexenal, whose peak area represented 38.99% of the total ion chromatogram. Unfortunately the chromatographic analysis could not distinguish between the *E* and *Z* isomers of 2-hexenal but their odor and taste thresholds are similar, with both showing an odor threshold of

0.01 ppm and taste thresholds of 0.06 and 0.08 ppm for the *E* and *Z* isomers, respectively (Hatanaka 1993). Although this compound is the most abundant, it is unlikely as in most fruits that the aroma of ovo derives principally from one compound, but rather from the complex interaction of the various aromatics present.

Economic Prospects

Between 1965 and 1994, the area devoted to the cultivation of ovo in Ecuador fluctuated between 400 and 1900 ha, with 1987 and 1990–1992 representing the only years in which more than 1800 ha were devoted to this fruit and in which more than 4500 metric tonnes were harvested (INEC 1987–1995, MAG 1994). The highest yield reported was 10 000 metric tonnes, which we estimate would be sufficient to make industrial processing of the fruit commercially viable (depending on the food product being produced, an industrial processor would probably need about 4500 metric tonnes of ovo per year). Current production levels, i.e., well below 4500 metric tonnes a year, necessarily limit commercial possibilities to processing ovo at the level of a cottage industry or small industrial scale.

Cottage Industry

In Ambuquí, 75% of the production of ovo is harvested green: these fruits ripen in 4–5 days and have a shelf life of an additional 2–3 days. Tree-ripened fruit has a shelf life of only 1–2 days after picking. As ovo is a soft fruit that is easily damaged during transport and as it has a limited shelf life the international exportation of ovo as a fresh fruit is rather restricted, especially given the scarcity of refrigerated and/or modified atmosphere storage facilities and shipping containers to delay ripening and prolong shelf life. Most of the production of ovo is destined for markets in the larger Ecuadorian cities, and although there are "local" reports of some exportation to Colombia and Perú (Macía 1997), no official register of such international trade exists. Its sale as a fresh fruit also limits the income of the small holders as the majority of the profits go to the middlemen who buy the fruit in bulk and transport it to the markets in the larger cities (Macía 1997). In Ambuquí, increases in tourism and a change to the cultivation of more profitable crops such as beans, tomatoes, and grapes combine to decrease the amount of land and effort dedicated to the cultivation of ovo.

To assure any economic future for ovo the small holders must concentrate on "value added products" rather than on the sale of fresh fruits. For example, small holders may be paid US$ 0.10–0.30/kg of fresh fruit (higher prices at the beginning of the season) whereas a 300 g jar of jam could be sold for US$ 1.45. With production costs[2] for such a jam roughly estimated at US$ 1.00/kg the small holders' profit increases to about US$ 3.83/kg of ovo processed. Distribution, however, still remains a problem and one must travel to the areas of production to purchase such products as the jam, ice cream and alcoholic beverages: marketing and advertising are virtually nonexistent. More aid from national and international sources is needed to help the cottage/small industry produce and promote ovo and ovo-based products.

In addition to the traditional products of jams, ice creams and alcoholic beverages, cottage production of an ovo vinegar, ovo flavored hot chili sauce, and dried fruit slices could be viable new products. In the coastal Province of Manabí (Ecuador) there is a cottage industry producing a banana vinegar which is unusual in that it results from a mixed acid fermentation. The best vinegars have acetic and lactic acids in a 2:1 ratio, which gives a pleasantly "rounded" acidity that complements the slight banana flavor of the vinegar. The high concentration of fermentable substrate in ovo, combined with its aroma, might yield an interesting fruit vinegar. Recently, a small food company has started marketing, and even exporting, hot chili sauces which are flavored with local fruits such as maracuyá (an acidic and aromatic passion fruit, *Passiflora edulis* var. *flavicarpa*). An ovo flavor might be an interesting addition to such a range of products.

---

[2] For producing a jam 33% ovo and 50% sugar by weight the production costs were calculated on the basis of: sugar purchased in bulk, US$ 0.40/kg; glass jars with metal lids purchased in bulk, US$ 0.21/each; and fuel US$ 0.10 to produce 1 kg of jam. The basic jam recipe was assumed to be 1 kg ovo, 1.5 kg sugar and 0.5 kg water. For 3 kg of jam: total production cost, US$ 3.00; sales US$ 14.50; profit US$ 11.50. Costs of fruit and labor were not included as these would be provided by the small holders. Overheads such as electricity and water were excluded as well.

Industrial

Being highly aromatic, ovo could probably be used industrially as an "exotic" fruit or flavoring for the extension of existing food product lines such as jams, candies and other sweets, ice creams, juices, and alcoholic beverages (perhaps most appropriately liqueurs as the unusually effervescent fermented beverage produced by the cottage industry has a tendency to cause bottles to explode). Ovo might even conceivably find a niche as an exotic fragrance in cosmetic products such as soaps, shampoos, hand creams, body lotions and other personal care products. Before ovo can be successfully commercialized on an industrial scale, food manufacturers will need to be convinced that its flavor is sufficiently "exotic" yet agreeable to consumers, that its use would contribute a competitive edge when extending product lines, and that sufficient supplies of the raw material of appropriate quality are available for processing.

Although an industrial interest in ovo would most likely provide the needed impetus for improvement of varieties in terms of yield, fruit to seed ratio, disease resistance, as well as improvements in methods of cultivation and postharvest management, the industry would not be likely to hold its interest in abeyance until such time that sufficient quantities of ovo of appropriate quality become available for processing. Nor would the small holders be likely to fare better economically by selling fruit to industrial processors rather than to the middlemen who currently market the fresh fruit.

Conclusions

Although ovo is valued as a seasonal treat, only its vitamin C content will have an important, albeit short-lived, nutritional impact on consumers. With few exceptions, yearly production has been under 4500 metric tonnes, implying that the per capita consumption of ovo in Ecuador is well below 650 g/year. It is uncertain to what extent local consumption could be expanded. International commercialization of ovo as a fresh fruit would require costly investment in improved post-harvest management (cold storage and/or modified atmosphere storage) and shipping facilities. Foreign purchasers would have to be assured of a reasonable shelf life for the sale of the fruit. A more promising economic future for ovo would derive from the promotion of cottage/small industry production of ovo-based products, which might eventually catch the interest of large-scale industrial food producers.

Acknowledgments

Gratefully acknowledged is the technical assistance with the chemical analyses provided by C. Espinoza, S. Fonseca, A. Proaño and Y. Sinche of the Nestlé R&D Center, Quito. To the small holders in Ambuquí we extend a very special thanks for their kind hospitality and willingness to share their information and experiences.

Literature Cited

Barfod, A. 1987. Anacardiaceae. Pages 9–49 *in* G. Harling and L. Andersson, eds., Flora of Ecuador, Vol. 30. Berlints, Arlöv, Sweden.

Boehringer Mannheim GmbH. 1989. Food analysis: starch uv method. Cat. No. 207748. Mannheim, Germany.

División de Investigaciones en Alimentos. 1983. Tabla de composición de alimentos para uso práctico. Publicación No. 42. Ministerio de Sanidad y Asistencia Social, Instituto Nacional de Nutrición, Caracas, Venezuela.

Duke, J. A., and A. A. Atchley. 1986. CRC Handbook of proximate analysis tables of higher plants. CRC Press Inc., Boca Raton, FL, USA.

Engel, K-H., J. Heidlas, and R. Tressl. 1990. The flavour of tropical fruits (banana, melon, pineapple). Pages 195–219 *in* I. D. Morton and A. J. MacLeod, eds., Food flavours. Part C. The flavour of fruits. Elsevier Science Publishers B. V., Amsterdam, The Netherlands.

Furia, T. E., and N. Bellanca (eds). 1975. Fenaroli's handbook of flavor ingredients. Vol. 2. 2nd ed. CRC Press, Inc., Cleveland, OH, USA.

Guthrie, H. A. 1979. Introductory nutrition. 4th ed. The C. V. Mosby Company, St. Louis, MO, USA.

Hatanaka, A. 1993. The biogeneration of green odour by green leaves. Phytochemistry 34(5):1201–1218.

INEC. 1987–1995. Encuesta de superficie y producción agropecuaria por muestreo de áreas. Instituto Nacional de Estadística y Censos (INEC), Quito, Ecuador. (Authors note: this survey is published yearly under the same title, so we have cited this reference only once giving the years of the publications from which we have extracted data.)

Instituto Nacional de Nutrición. 1965. Tabla de composición de los alimentos ecuatorianos. Ministerio de Previsión Social y Sanidad, Quito, Ecuador.

Kochhar, S. P. 1996. Oxidative pathways to the formation of off-flavours. Pages 168–225 *in* M. J. Saxby, ed., Food taints and off-flavours. 2nd ed. Blackie Academic & Professional, London, UK.

León, J., and P. E. Shaw. 1990. Spondias: the red mombin and related fruits. Pages 116–126 *in* S. Nagy, P. E. Shaw, and F. W. Wardowsky, eds.,

Fruits of tropical and subtropical origin. Florida Science Source, Lake Alfred, FL, USA.

Leung, W-T.W., Butrum, R. R., and Chang, C. F. 1972. Proximate composition, mineral and vitamin contents of East Asian Food, Part 1, *in* Food and Agricultural Organization and U.S. Department of Health, Education and Welfare, Food composition Table for use in East Asia, Washington DC, USA.

———, and M. Flores. 1961. Tabla de composición de alimentos para uso en América latina. National Institutes of Health, Bethesda, MD, USA.

McCance, R. A., and E. M. Widdowson. 1991. The composition of foods. 5th ed. The Royal Society of Chemistry and the Ministry of Agriculture, Fisheries and Food, London, England.

McCready, R. M., and E. A. McComb. 1952. Extraction and determination of total pectic materials in fruits. Analytical Chemistry 24:1986–1988.

Maarse, H., and R. E. Kepner. 1970. Changes in composition of volatile terpenes in Douglas fir needles during maturation. Journal of Agricultural and Food Chemistry 18:1095–1101.

Macía, M. J. 1997. El "ovo" (*Spondias purpurea* L., Anacardiaceae) un árbol frutal con posibilidades socioeconómicas en Ecuador. Pages 271–281 *in* M. Ríos and H. B. Pedersen, eds., Uso y manejo de recursos vegetales: Memorias del segundo simposio ecuatoriano de etnobotánica y botánica económica. Abya-Yala, Quito, Ecuador.

MAG. 1994. Primer compendio estadístico agropecuario del Ecuador 1965–1993. Ministerio de Agricultura y Ganadería (MAG), Quito, Ecuador.

Merrill, A. L., and B. K. Watt. 1973. Energy value of foods: basis and derivation. Agricultural Handbook No. 74. United States Department of Agriculture, Washington, DC, USA.

Mitchell, J. R., K. Buckley, and I. E. Burrows. 1978. Food binding agent. UK Patent 1 525 123. The Patent Office, London, England.

Morton, J. F. 1987. Fruits of warm climates. Julia F. Morton, Miami, Florida, U.S.A.

National Research Council. 1989. Recommended dietary allowances. 10th ed. National Academy Press, Washington, DC, USA.

Nickerson, G. B., and S. T. Likens. 1966. Gas chromatographic evidence for the occurrence of hop oil components in beer. Journal of Chromatography 21:1–5.

Shibamoto, T., and C-S. Tang. 1990. 'Minor' tropical fruits—mango, papaya, passion fruit, and guava. Pages 221–280 *in* I. D. Morton and A. J. MacLeod, eds., Food flavours. Part C. The flavour of fruits. Elsevier Science Publishers B. V., Amsterdam, The Netherlands.

Souci, S. W., W. Fachmann, and H. Kraut. 1986. Food composition and nutrition tables 1986/87. Wissenschaftliche Verlagsgesellschaft mbH, Stuttgart, Germany.

Sturrock, D. 1959. Fruits for southern Florida. Southeastern Printing Co., Stuart, Florida, USA.

Winton, A. L., and K. B. Winton. 1935. The structure and composition of foods. Volume II: Vegetables, legumes, fruits. John Wiley & Sons, Inc., New York, U.S.A.

Uses of Saw Palmetto (*Serenoa repens*, Arecaceae) in Florida[1]

Bradley C. Bennett and Judith R. Hicklin

Bennett, Bradley C. (*Departments of Biological Sciences & Environmental Studies, Florida International University and Fairchild Tropical Gardens, Miami, FL 33199*) **and Judith R. Hicklin** (*Department of Biological Sciences, Florida Intentional University, Miami, Florida 33199*). Uses of Saw Palmetto (*Serenoa repens*, Arecaceae) in Florida. *Economic Botany, 52(4):381–393. 1998. Saw palmetto* [Serenoa repens (*Bartr.*) *Small*] *is the most common native palm in the U.S. and, possibly, the most useful. Cattle and wildlife, especially black bears and white-tailed deer, consume the fruits. The flowers are a favorite nectar source for honey bees, and the sprawling, shrubby palm provides excellent cover for birds, reptiles, and small animals. Saw palmetto's edible fruits were a staple in the diet of Florida's pre-contact inhabitants. Vegetative parts of the plant supply fiber, wax, and roof thatch. Current interest in saw palmetto stems from its use in the treatment of benign prostatic hyperplasia. An estimated 6 800 000 kg of the fruit are shipped to Europe each year. Preliminary data suggest that the economic value of saw palmetto harvesting may exceed that of cattle grazing, the common use of much saw palmetto habitat.*

La Utilización de Saw Palmetto (*Serenoa repens*, Arecaceae) en el Estado de Florida USA. *Saw palmetto* (Serenoa repens) *es la palma mas común en los Estados Unidos y, quizás se puede decir, es la mas utíl. El ganado y los animales silvestres, particularmente los osos negros y los venados cola-blanca, comen las frutas. Las flores del saw palmetto son una fuente favorita de néctar para las abejas de miel, y la palma extendido y fruticoso ofrece una cubierta excelente para los reptiles, aves, y animales pequenos. Las frutas comestibles del saw palmetto fueron un alimento principal en la dieta de la gente indígena de la Florida. Las partes vegetativas producen fibra, cera, y hojas para hacer techos. Hoy en día, el interés en saw palmetto resulta del su uso para tratar la inflamacíon benigna de la glándula próstata. Se estima que una cantidad de 6 800 000 kg de las frutas es exportada a Europa cada año. Estudios preliminares sugieren que el valor económico de la cosecha de saw palmetto excede el valor de ganadería, cual es el uso mas común de la tierra donde crece saw plametto.*

Key Words: saw palmetto; *Serenoa repens*; benign prostatic hyperplasia; prostate; Seminole; Miccosukee; Florida; plant medicines; plant fibers; Arecaceae; non-timber forest products.

Saw palmetto (*Serenoa repens* (Bartr.) Small), the most common palm in the U.S., grows throughout the state of Florida. Its native range extends north through the coastal plain of Georgia and southern South Carolina and west through the coastal plain of Alabama and Mississippi (Small 1926). Although some consider it to be a bothersome rangeland weed, this abundant palm provides wildlife forage and an array of products for humans, including fiber, wax, and medicines. In this paper, we discuss current and historical uses of *Serenoa repens*, especially its medicinal value in the treatment of benign prostatic hyperplasia.

[1] Received 26 March 1997; accepted 10 April 1998.

Taxonomy

Serenoa repens is the sole representative of *Serenoa* (Arecaceae, subfamily Coryphoideae, tribe Coryphae, and subtribe Livistoninae). Closely related genera include *Brahea* and *Acoelorraphe* (Uhl and Dransfield 1987), the latter of which also grows in Florida. The pharmacological and medicinal literature often cites the name *Serenoa serrulata* (Michaux) Nicholson. Other synonyms include *Brahea serrulata* (Michaux) H. Wendl., *Chamaerops serrulata* Michaux, *Corphya obliqua* Bartram, *Corypha repens* Bartram, *Diglosophyllum serrulatum* (Michaux) Wendl. ex Drude, and *Sabal serrulata* (Michaux) Nuttall (Bailey 1945; Glassman 1972; Zona 1997).

Economic Botany 52(4) pp. 381–393. 1998

J. D. Hooker established the genus *Serenoa* to honor Sereno Watson (1826–1892), an introverted assistant of Harvard's Asa Gray (Feurt 1953; Zona 1997). The specific epithet, repens, refers to the plant's creeping habit. *Serenoa repens* aptly is called saw palmetto throughout its range. Petiole margins are well armed with abundant saw-like teeth; palmetto is a Spanish derivative meaning small palm.

Description

Saw palmetto stems are subterranean or prostrate, though they occasionally ascend obliquely to a height of 2 m or more. The stems are unusual among the palms in that they frequently branch. The fan-shaped leaves achieve a width of 1 m and may be divided more than half of their length. Leaves emerge from a fibrous brownish sheath. Two vegetative forms occur—the common yellow-green type and a less common glaucous blue-green type (Zona 1997). Inflorescences emerge from February to mid-April (Hilmon 1968). They are about as long as the foliage and have as many as 3–4 orders of branching (Uhl and Dranfield 1987). Fertile ramets produce between one and five inflorescences. Both the tubular calyx and cream-colored corolla are 3-lobed. Sepals are 1–2 mm long and petals are 3–5 mm long. Flowers bear 6 stamens, a tricarpellate ovary, and a single style. A single fertilized carpel develops into a black drupaceous fruit with a thin endocarp. The 2 cm long drupes ripen in September and October (Long and Lakela 1976; Radford et al. 1968; Uhl and Dransfield 1987).

In most clustering palms, branches arise from buds located at or below ground level. As the main axis matures, there is a switch from vegetative to reproductive branching. Clustering, therefore, is restricted to the basal part of the stems. Fisher and Tomlinson (1973) described a contrasting growth pattern in *Serenoa,* which lacks a clear distinction between vegetative branching and inflorescence development. They note that about 50% of axillary buds abort, 40% become inflorescences, and the remaining 10% become vegetative suckers. The alternation between vegetative and reproductive branching is irregular. Saw palmetto thus may form dense clumps arising from underground stems or more diffuse clumps that arise from above ground vegetative shoots.

Fig. 1. a) Saw palmetto forming the shrub layer in a pine flatwood community in Martin County, FL., b) Saw palmetto growing on exposed Miami Limestone in a pine rockland community in Dade County, FL, c) Saw palmetto forming part of the shrub layer in a coastal pine scrub community in Palm Beach County, FL.

Ecological Distribution

One of the most abundant plants in Florida, saw palmetto is a characteristic element of pine flatwoods, prairies, scrub, and mesic hammocks (Fig. 1). It also is a common component of dunes and maritime forest, but may be sensitive to salt spray (Abrahamson and Hartnett 1990; Johnson and Barbour 1990; Kushlan 1990; Myers 1990; Platt and Schwartz 1990). All of these communities are well-adapted to fire and saw palmetto is one of the most fire-tolerant of the pyric species (Fig. 2 and 3). Craighead (1971) described the formation of palmetto rings after

Fig. 2. Burning saw palmetto during a prescribed burn in Martin County, FL coastal pine scrub community.

severe fires destroyed the interior hardwood hammock trees in the Everglades. The plant also grows in short-hydroperiod swamps and along the edges of tropical hammocks in southern Florida.

Following WWII, saw palmetto coverage was estimated to be 1 420 000 hectares or about 10% of Florida's land surface (Anonymous 1947). Saw palmetto often forms the shrub layer in *Pinus elliottii* (slash pine), *P. serotina* (pond pine), and *P. palustris* (longleaf pine) flatwoods. Though many flatwoods have been converted to rangeland and pine plantations, they still abound in the southwestern and east coast lowlands, northern Gulf Coast, and north-central peninsular regions (Abrahamson and Hartnett 1990).

Wildlife Use and Ecological Value

Raccoons, foxes, rats, gopher tortoises, whitetail deer, black bear, and feral hogs feed extensively on saw palmetto drupes; fish and waterfowl consume them as well (Abrahamson and Abrahamson 1989; Hale 1898; Smith 1972). The fruits are an especially important late-fall and early-winter food for whitetail deer (Elias 1980). John Lloyd, an early American botanist, observed that animals that fed on the fruits "grew sleek and fat" (Ritchason 1994). Feurt noted (1953) that, "Cows eating saw palmetto berries give more and richer milk than on any other kind of food." Others have commented on the fruit's wildlife value. Hale (1898) described the, "great fattening properties of the berries, and the peculiar quality of the fat of the animals that fed on them."

Cattle consume the leaves only when better forage is lacking (Wilder and Kitzke 1954). For that reason, ranchers have led the assault on saw palmetto in Florida. In Europe, oil cakes for livestock are made from the fruit and seed residue. Animals reportedly fatten more rapidly on these than on rape or flax seed (Sherman and Briggs 1899 cited in Feurt 1953). *Serenoa* flowers are a good honey source (Duke 1985; Wilder and Kitzke 1954) and they also attract butterflies (Tasker 1994). Besides forage, the plants provide excellent coverage for wildlife, including the endangered Florida panther (*Felis concolor* ssp. *coryi*) and the threatened Florida black bear (*Ursus americanus* ssp. *floridanus*). Black bears often give birth in the protective cover of saw palmetto thickets. The plant provides food or cover for more than 100 bird, 27 mammal, 25 amphibian, and 61 reptile species (Maehr and Layne 1996).

Fig. 3. Recently burned saw palmetto stems in a beach strand community in Palm Beach County, FL.

Human Use

Food

For humans, saw palmetto is one of Florida's most versatile plants, providing food, fiber, oil, medicine, wax, and roof thatch. With cocoplum (*Chrysobalanus icaco* L.) and sea grapes [*Coccoloba uvifera* (L.) L.], saw palmetto is ubiquitous in coastal archaeological sites throughout southern and central Florida. It was one of the most important foods for Florida's pre-Columbian, non-horticultural peoples and later Creek immigrants (e.g., McGoun 1993; Reitz and Scarry 1985; Widmer 1988). Saw palmetto also was one of the wild foods found in settlements from sixteenth century Spanish Florida (Reitz and Scarry 1985). In the early 1900s, Miami pioneers mixed juice from saw palmetto berries with carbonated water and sold the product as a soft drink called "metto" (Morton 1977). Not all European descendants, however, viewed the fruits with favor. Jonathan Dickinson, a young Quaker merchant, and a party of 20 were shipwrecked August 23, 1696 on Florida's east coast. Held captive by the Ais and Jaega for several weeks, Dickinson recorded his oft-cited impression of saw palmetto fruits (Dickinson 1945).

> "We tasted them, but not one amongst us could suffer them to stay in our mouths; for we could compare the taste of them to nothing else, but rotten cheese steeped in tobacco."
>
> "In some time after we had been in the house, came in Indian women laden with baskets of berries, mostly of the palm, some seaside coco-plums and seaside grapes: of the two latter we could eat; but of the palm-berries we could not bear the taste in our mouths."
>
> "The Indians offered us some of their berries, which we endeavored to eat; but could not; the taste was so irksome and ready to take our breath from us, when we tried to eat them."

The plant's infamous fruits have even found their way into fiction. A character in Robert Wilder's *God has a Long Face* responded concisely but colorfully to a tonic made from saw palmetto, "Tastes like a mouth full of cockroaches" (Wilder 1940 cited in Morton 1986). In contrast to the contempt of Europeans, Florida's original people cherished the fruit, as Dickinson described a cacique "greedily eating" the fruits. Florida's Seminole and Miccosukee peoples, whose ancestors migrated to Florida after the extermination of the pre-Columbian inhabitants, also consumed the fruits. Among the contemporary population, some older Seminoles still eat them, although the younger ones share Dickinson's disdain. Even saw palmetto connoisseurs eat the fruits in moderation. A Seminole man warned, "You can't eat too many fruits, because they are bitter and after eating them hot liquids burn the mouth. Another Seminole said, "You should not eat more than five fruits at a time."

The Seminoles also consume the fruits in a drink called "shiope sofkee." They squeeze the fruits to collect the juice, then add sugar before drinking it. In historic times, particularly during the Seminole wars, saw palmetto stems were an important starch source. The stems were chopped, ground, then mixed with water and strained. The resulting flour was used in sofkee (Bennett ms.). As with the sympatric *Sabal palmetto* (Walt.) Lodd. ex J. A. & J. H. Schultes, saw palmetto's vegetative buds also are edible (Small 1926).

Saw palmetto fruits are rich in oils, which often are lacking in indigenous diets. Nearly two-thirds of the oil is composed of free fatty acids including capric, caprylic, caproic, lauric, palmitic, and oleic acids (Murray 1992). Abrahamson and Abrahamson (1989) found that the fruits contained more fat, digestible carbohydrates, and digestible nutrients than those of the sympatric *Sabal etonia* Swingle ex Nash.

Fiber

Saw palmetto fibers, obtained from the leaf sheaths, stems, and roots, have been used widely. Scrubbing brushes readily are made from the tough fibrous roots, which are similar in texture to coconut husks. The roots also provide material for the manufacture of paper (Hale 1898) but of poor quality (Wilder and Kitzke 1954). Leaves and stem fibers have been used commercially as upholstery fill, brush fiber, and plaster reinforcer. During WWII, a factory was erected in Florida to use the root pulp in cartridge plugs and for wall board (Robinson 1947). An estimated 600 kg of fiber cork could be obtained from 1 metric ton of palmetto roots (Anonymous 1947). The stem once was processed into an acceptable cork substitute for floats and gun plugs (Wilder and Kitzke 1954), and stem fibers provided material for ropes and rugs.

Serenoa repens, known as *sheope'^{n} taale* among Hitchiti-speaking Miccosukees and Seminoles and *shiopa mahi* or *shiapo entaale* among Creek-speaking Seminoles (Bennett pers. obs.; Sturtevant 1955), is an important fiber source for

Fig. 4. a) Saw palmetto petiole fibers, b) Seminole doll with a face made from saw palmetto petiole fibers.

Florida's extant indigenous people. The ancestors of Florida's extant native people began crafting baskets from saw palmetto petioles, instead of the more traditional split cane, as they migrated south during the 18th and 19th centuries. Deaconess Harriet Bedell rekindled interest in basket making in the 1930s when she began encouraging Seminole women to make tourist baskets from sweet grass and palmetto (Downs 1995).

After drying the petioles, the Seminoles and Miccosukees weave winnowing and fruit-collecting baskets from adaxial strips and fashion sieves from abaxial strips. Palmetto baskets have more than utilitarian value. Corn sifting baskets are important during the annual Green Corn Dance and, because of their ritual significance, they traditionally were made only by married women. A set of ritual baskets consists of two open-weave containers, which are used to sift the ground corn, and a third of closed weave, which holds the sifted meal (Downs 1995). Saw palmetto baskets also are used to hold medicinal plants (Bennett, ms.).

Although the coarse palmetto fibers are difficult to collect and process, both Neill (1956) and Downs (1995) cited the frequent use of the fibers in Seminole and Miccosukee crafts. Downs employed the common name palmetto for both craft and construction materials; the latter are derived from the cabbage palm (*Sabal palmetto*). The famous Seminole dolls are made exclusively from *Serenoa repens* petiole fibers, which are prized because of their reddish colors (Fig. 4). Other indigenous fiber uses include the manufacture of brushes, skimmers, rope, tinder, and tourist goods (Sturtevant 1955).

Construction

The leaves of *Sabal palmetto* are the preferred thatch material in Florida, but *Serenoa repens* leaves also are used occasionally. Hale (1898) reported the use of *Serenoa* leaves for roof thatch. The open-sided, Seminole chickees, especially small temporary structures, may be thatched with saw palmetto leaves. There are frequent references to the use of palmetto thatch in both indigenous and pioneer housing, but authors seldom distinguish between *Serenoa repens* and *Sabal palmetto* (e. g., Dickinson 1945; Pierce 1970).

WAX SOURCE

Palmetto leaves, especially the blue-green form, are extremely waxy. The hard, brown, resinous wax has an acid number of 16.3, saponification number of 101.5, melting point of 81.2°C, and iodine number of 9.8. Free flaking yields range from 1.4 to 2.7 grams per 100 grams of leaf, while total yield averages 5.8 grams for the same quantity of raw material. Wilder and Kitzke (1954) noted that this is "a surprisingly high yield nearly equivalent to carnuba." Unfortunately, most commercial palm wax is the easily-collected free flaking type. The cost of harvesting and processing limits saw palmetto's commercial value as a wax source.

ORNAMENTAL

Benefitting from renewed interest in Florida's native plants, saw palmetto now is a widely planted ornamental. Simpson (1926) advocated the use of the plant in low areas where, if protected from fire, it "becomes almost a tree . . . , as it produces a startling effect." Bush and Morton (1963) called saw palmetto an ideal shrub for highway plantings because it requires no maintenance and controls erosion on steep embankments. Craig (1991) listed the palm as a suitable plant for coastal dunes, especially in the back dune zone. Saw palmetto's landscape potential is somewhat limited by its difficulty in transplanting and slow growth. Material is not commonly available in nurseries and often is more expensive than easily propagated exotics. Enlightened landscapers, however, incorporate saw palmetto into their designs.

MISCELLANEOUS

Humans employ saw palmetto for a variety of other purposes. Florida's pre-Columbian people fashioned plate substitutes and fire fans from the leaves (Dickinson 1945) and roasted food on the sharpened petiole (Swanton 1922). The Miccosukee and Seminoles make ceremonial dance fans, fire fans, and baby rattles from the foliage (Bennett ms; Sturtevant 1955). The latter were constructed by weaving a small basket from the leaf fibers while they are still attached to the petiole. A piece of charcoal was inserted in the basket to produce the rattling sound. The Seminoles also use saw palmetto leaves for bedding in temporary camps.

Serenoa repens fruits and seeds are the potential source of a commercial volatile oil and fixed oil. The roots also were investigated as a source of tannic acid. Researchers estimated that the production of saw palmetto fibers for wallboards would yield 417 liters of a tannic acid solution from 1 metric ton of roots (Anonymous 1947). The fruits also have been used as an aromatic in cognac (Duke 1985).

MEDICINE

Of saw palmetto's many products, remedies made from the fruits count among the most significant (Table 1). Farnsworth and Soejarto (1985) ranked saw palmetto with the most useful plant-derived drugs in the U.S., and it ranked among the top ten selling herbs from a survey of selected health food stores (Brevoort 1995). The fruits reportedly possess anticatarrhal, antitumor, aphrodisiac, digestant, diuretic, narcotic, and sedative properties (Grieve 1971; Hartwell 1970; Maisch 1899; Sargent 1899; Sollmann 1917). They have been used to treat renal disorders, hypertrophy of the prostate, cystitis, orchitis, diarrhea, dysuria, epididymitis, and bronchitis and to stimulate breast growth (Feurt 1953; Madaus 1938 cited in Feurt 1953). Combined with other herbs, saw palmetto fruits reputedly stimulate the appetite, improve digestion, and increases assimilation. As a nutritive tonic, the fruit is said to increase the size and secreting ability of the mammary gland and decrease ovarian and uterine irritability. *Serenoa* are said to stimulate bladder flow and are employed to treat male importence, sterility, and reproductive problems. These extracts are reported to increase blood flow to the sexual organs and affect the thyroid gland's regulation of sexual development. Fruit preparations are also said to "quiet the nerves," decrease nasal and throat secretions, and to have antiseptic properties (Ritchason 1994).

Saw palmetto was listed in the United States Pharmacopoeia from 1900 to 1920 and the National Formulary from 1926 to 1952 (Feurt 1953). Official preparations are listed in Table 2. The Extra Pharmacopoeia of Great Britain

TABLE 1. MEDICINAL USES OF SAW PALMETTO.

Condition or use		
Alchoholism[5]	Dysuria[2,3,4]	Ozena[3]
Alopecia[3]	Epididymitis[1,2,4]	Parasiticide[5]
Anemia[3]	Expectorant[1,2,5]	Prostate gland (enlargement)[1,3,4,5,6]
Anticatarrhal/ antiinflammatory[1,3,5,6]	Genito-urinary stimulant[1,5]	Reproductive organs[3,5]
Antiseptic[5]	Gonorrhea[3]	Respiratory congestion & infections[1,3,5]
Antispasmodic[5]	Headaches[3]	Rheumatism[3]
Antitumor[3]	Hemorrhoids[3]	Sedative[1,2,5]
Appetite stimulant[3,5]	Hormone regulation[5]	Sexual stimulant[1,2,3,5,6]
Asthma[2,3,5]	Impotence, male[3,5]	Sore throat[5]
Breast atrophy[1,3,6]	Incontinence, urinary[3]	Sperm production[6]
Breast enlargement[1,2,3,4,5,6]	Infertility[5]	Stomachache[1]
Bronchitis[1,2,3,4,5]	Iritis[3]	Testicular atrophy[1,3,6]
Colds[1,5]	Kidney disease (incl. Bright's)[2,5]	Tonic[1]
Diabetes[1,5]	Laryngitis[3]	Urinary tract infection & cystititis[1,2,3,4,5,6]
Diarrhea & Dysentery[1,3,4]	Leukorrhea[3]	Weight gain, promotion[5]
Digestive aid[1,3]	Neuralgia[3,5]	Whooping cough[3,5]
Diuretic[1,2,6]	Nutritive[1]	
Dysmenorrhea[3]	Obesity[5]	
	Ovarian atrophy[3]	

[1] Duke (1985), [2] Feurt (1953), [3] Hale (1898), [4] Maddaus (1938) cited in Feurt (1953), [5] Ritchason (1994), [6] Tyler (1982).

and the British Pharmaceutical Codex sanctioned saw palmetto's use for treating cystitis, chronic bronchitis, and for its stimulation of the genito-urinary tract and mucous membranes (Feurt 1953). The fruits are still listed in European pharmacopoeias (M. Huffman, personal communication; Tyler 1982, 1994); the United States Pharmacopoeia again is considering saw palmetto, this time for its list of approved dietary supplements (W. Popin, personal communication). Feurt (1953), in a pharmacological study of the fruits, showed that their vasoconstrictor effects accounted for their use as stimulants, vitalizers, nasal decongestants, and diuretics.

Several unofficial *Serenoa* preparations, once available in the U.S. (Hale 1898), were administered to treat genito-urinary diseases; inflammation of pulmonary, bronchial, and laryngeal membranes; and anemia (Table 3). Saw palmetto products are still available in U.S. health food stores, where they are sold as dietary supplements (Fig. 5a). Body builders take saw palmetto dietary supplements with the intent of increasing muscle mass. The fruits also are ingredients in several products sold as aphrodisiacs Yohimbe Plus®, manufactured by Action Labs, contains 100 mg of a saw palmetto extract and Cobra Performance Enhancer® manufactured by Pep Companies, Inc., contains an unspecified amount of saw palmetto.

Serenoa fruits are most valued for treating prostate problems (Duke 1985; Tyler 1982, 1994). In a letter to Sir William Thiselton-Dyer of the Royal Botanic Gardens Kew, C. S. Sargent wrote about saw palmetto's efficacy, "Its medical values are highly prized for the treatment of all diseases of the mucous membrane, and especially for the alleviation of troubles of the prostate gland" (Sargent 1899). Permixon, Talso, Strogen Forte, Prostagutt, and Remigeron, all European drugs manufactured from dried saw palmetto fruits, are reported to be highly effective in the treatment of benign prostatic hyperplasia (BPH). Oral doses of a liposterolic saw palmetto extract (160 mg, twice daily) increased urinary flow rates and volume and decreased prostate size after 45 days of treatment. In a clinical trial involving 505 patients, 88% of both the physicians and the patients considered the treatment to be effective. Side effects were reported in only 5% of the patients, a rate that compares favorably with those of existing therapies (Braeckman 1994; Bruhwyler 1994).

Dried saw palmetto fruits contain high concentrations of sitosterols. Scientists originally believed that estrogenic activities of the sitosterols were responsible for the fruits' therapeutic

TABLE 2. HISTORIC AND CURRENT PREPARATIONS OF SAW PALMETTO FRUITS.

Type of preparations	Description
Historic (from Hale 1898).	
Tincture & dilutions	macerated fresh fruits placed in 90% alcohol for 14 days, then decanted.
Fluid extract	4 times stronger than tincture, can be combined with syrups or maltine.
Oil	dilutions made from saw palmetto oil used in tablets or discs.
Saccharated oil	1 part saw palmetto oil to 7 parts cane sugar.
Malto-Sabal	1 part saw palmetto oil to 7 parts maltine or maltose.
Aqua Oleum Sabal	saw palmetto oil, magnesium carbonate, and distilled water, for use as a spray.
Suppositories	5–10 drops of saw palmetto oil with cocoa butter.
Current (from W. Popin, personal communication)	
Teas	—
Powders, traditional	in 2 piece hard gelatin capsules.
Powder, extracts	25–30% fatty acid content.
Liposterols extracts	in soft gelatin capsules.
Standardized extracts	standardized to a specific fatty acid or liposterol content.

effectiveness. Experimental work now suggests that saw palmetto's effects on the prostate may be due, in part, to its antiandrogenic activity. The constituents responsible have not been identified, but they are known to occur in the fat-soluble fraction of the fruit extract. Using electrospray mass spectrometry, J. M. E. Quirke, an analytical organic chemist working with us, has isolated four previously unreported fat-soluble compounds from saw palmetto fruits. We are now characterizing these compounds.

Liposterolic fruit extracts reduce the uptake of testosterone and dihydrotestosterone (DHT) by more than 40%. DHT binds to cytoplasmic receptors in the prostate, where it initiates RNA

TABLE 3. UNOFFICIAL PREPARATIONS (FROM HALE 1898).

Preparations	Description
Sanmetto	saw palmetto oil, sandalwood oil (***Santalum album*** L.), and *Triticum repens* [sic].
Saw palmetto compound	saw palmetto, cola (*Cola* spp.), coca (*Erythroxylum coca* Lam.) and parsley seeds (*Petroselenium crispum* (Miller) A. W. Hill.
Sabal Palmetto & Santal Compound	saw palmetto, corn silk (*Zea mays* L.), and sandalwood oil.
Syrup of oil of saw palmetto with hypophosphates	saw palmetto oil and hypophosphates of lime and soda.

and DNA synthesis. This action, in turn, induces protein synthesis and abnormal growth of the prostate gland (Tyler 1994). Extracts also reduce conversion of testosterone to DHT by inhibiting the enzyme steroid 5α-reductase. The anti-inflammatory properties of *Serenoa* also may be responsible for its efficacy in relieving prostate swelling (Tyler 1982).

Fig. 5. a) Saw palmetto products available in U.S. health food stores, b) Saw palmetto cooperative in Immokalee, FL (Collier County).

Fig. 6. a) Saw palmetto infructescence bearing immature fruits, b) Saw palmetto fruits drying at Plantation Medicinals, Hendry County, FL.

Despite its importance as a North American botanical and a European drug, there are no reports that Florida's indigenous Seminoles or Miccosukees employ the plant medicinally. Medicine men make blowing tubes by tying two saw palmetto petioles together (a traditional medicine is believed to have power only after the medicine man breathed his spirit into it through a blowing tube). The fruits, however, have no place in the Seminole Pharmacopoeia.

Commercial Production

Several established botanical companies ship whole, dried palmetto fruits to Europe and new businesses have been established recently due to growing interest in the fruits. The U.S. market consumes a smaller but not insignificant quantity of the fruits. The Upjohn Company operated a processing plant in central Florida in the early 1900s (M. Huffman, personal communication). Today, fruit collecting is centered farther south around Immakolee, with secondary centers in Naples (Collier County), Fellsmere (Indian River County), and Winter Haven (Polk County). Johnson (1960) reported that collection was once localized around Cape Canaveral (Brevard County), and in the early 1900s, fruit gathering was centered around New Smyrna (Volusia County, M. Huffman, personal communication). Average annual fruit harvest is unknown, but may exceed 6 800 000 kg per year. Seasonal workers, often Spanish-speaking immigrants, collect the unripe fruits (Fig. 6a) and sell them to intermediaries (Fig. 5b). Although much fruit is harvested on private lands, most of Florida's county, state, and national parks have reported saw palmetto poaching (Benshoff 1995). The intermediaries sell the fruits to one of the several botanical companies located in the state. The major botanical exporters often establish contracts with large landowners for a fixed quantity of fruits at a fixed price.

Fruits, of varying degrees of maturity, arrive at the botanical companies in large crates that

Fig. 7. a) Saw palmetto fruits as they arrive in shipping crates at Plantation Medicinals, Hendry County, FL., b) Dried and cleaned sacks of saw palmetto fruits awaiting shipment to Europe at Plantation Medicinals, Hendry County, FL.

also are used for shipping oranges (Fig. 7a). Fully matured black or partially-ripened orange fruits are preferred but buyers permit a percentage of green fruits in their purchases. The fruits are then dried, often a considerable expense (Fig. 6b). During the season's peak, Plantation Medicinals, the leading U.S. exporter, burns $10 000 worth of propane each day in its driers (M. Huffman, personal communication). After they are dried, the fruits are mechanically cleaned then packed into sacks for shipment (Fig. 7b).

Saw palmetto fruits have been exported to Europe since 1602 (M. Huffman, personal communication). The European Pharmacopoeia requires whole, dried fruits and mandates that the fat-soluble constituents be included in the derived drugs. Organic solvents, including ethyl alcohol and hexane, or supercritical CO_2 are used to extract the liposterols. Hydraulic presses also have been used to process alcohol-preserved fruits. Additional oil has been recovered by mixing the residual seeds and pericarp with warm water, then skimming the oil from the surface (Sherman and Briggs 1899 cited in Feurt 1953).

Economic Potential

In a typical season, migrant workers receive $0.22–$0.44 for each kg of fresh fruits that they harvest. Due to low production, increased demand, and competition among European buyers, prices exploded to a high of $7.20 and an average of $4.85 per kg in 1995. Prices returned to a more normal average of $0.27 per kg in 1996 (Fig. 8). In 1961, a Georgia buyer offered the same price for large quantities of the dried fruits (Hilmon 1968), which are about one third the weight of fresh fruits. Brevoort (1995) reported wholesale prices at $48.56 per kg.

Fruit yields are estimated to range from 500 to more than 3000 kg per hectare per year on managed rangeland. Using the conservative values of 750 kg and $0.33 per kg (average price between from 1989 through 1996, discounting the abnormally high 1995 values), average gross yield per hectare is about $250 per hectare per year. This compares favorably to alternative land uses.

Several management techniques improve saw palmetto yield. Fruit production peaks 2–3 years after prescribed fires (Hilmon 1968). Our initial field observations suggest that lack of light may limit fruiting. Exposed saw palmetto clumps produce more inflorescences and have higher fruit set per inflorescence than clumps that occur in dense stands. Abrahamson and Abrahamson (1989) found that *Serenoa* fruits increased in size and fresh mass as moisture levels increased. They also noted that crude fat and fiber in the exocarp and mesocarp decreased with increased pH and improved soil drainage.

Hale (1898) noted that fruit is most abundant in alternate years, an observation collaborated by other observers (M. Huffman, personal communication). Whether this variation is synchronized across the geographic range of saw palmetto is unknown. The economic sustainability of saw palmetto production, depends, in part, on relatively low fruit prices ($0.33 to $0.55 per kg). At higher prices, European markets may shift to *Prunus africana,* which also is effective in the treatment of BPH (Cunningham and Mbenkum 1993).

Saw palmetto coverage in Florida is unknown, but assuming that the post WWII estimate of 1 420 000 hectares was accurate and that one third of that area currently is harvestable, there may be more than 450 000 hectares avail-

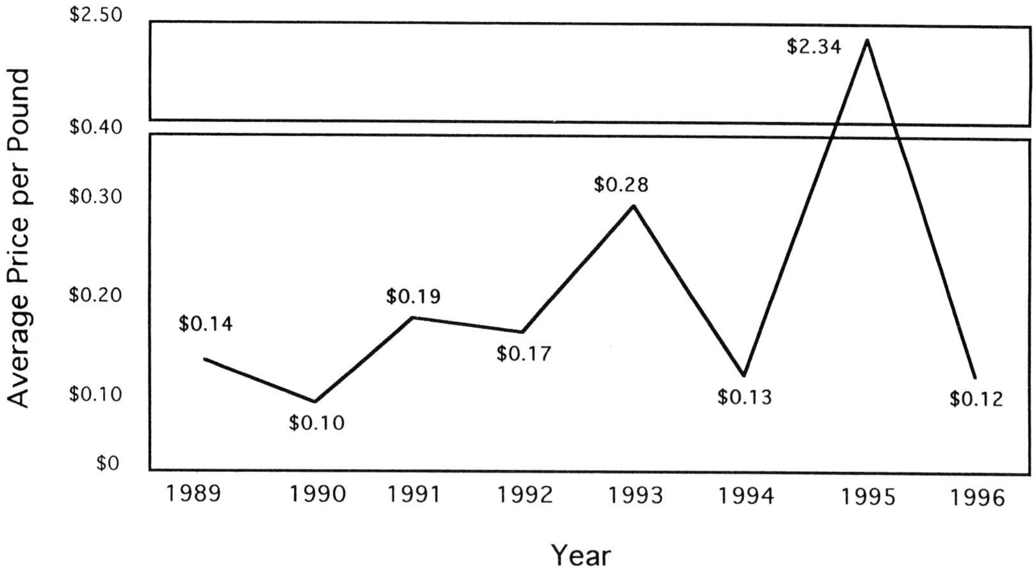

Fig. 8. Average saw palmetto prices (per pound) paid to collectors from 1989–1996.

able for harvest in the U.S. Further assuming a conservative harvest of 100 kg per hectare and a static demand of 6 800 000 per year, then 68 000 hectares would supply the needed saw palmetto fruits. Even a 3-fold increase in demand, far higher than any exporters foresee, could be met in 204 000 hectares or less than one half of the land available for harvest.

Collecting palmetto fruits on private land or exporting the dried material currently is unregulated. Because of its abundance and widespread distribution, existing harvest levels may have negligible effects on saw palmetto recruitment or the organisms that feed upon it. Maehr and Layne (1996), however, suggested that overharvesting in some regions may expose black bears to greater risks as they are forced to forage more widely. Except when prices are very high (e.g., the aberrant year 1995), collectors are unlikely to harvest all the fruits within a stand. It is easier to locate additional stands than to cut through a palmetto thicket. Dense saw palmetto stands often harbor eastern diamondback rattlesnakes, and several collectors reportedly have been bitten during the past two years. Harvesting along roadsides and trails is easier and safer. Collection, however, is highly localized and the effects may be severe during years of low fruit production, on accessible public lands, and in areas close to buyers. Increasing prices also may amplify collection rates. Intensive harvesting when prices are high (and mast production is low) will have significant localized effects on wildlife and palmetto regeneration. Several agencies including the U.S. Fish and Wildlife, Florida Department of Environmental Protection, Florida Game and Freshwater Fish Commission, and the U.S. Department of Agriculture have expressed concern over excessive harvesting. This concern may well result in future restrictions.

Conclusions

Saw palmetto provides food and habitat for many species of wildlife. It supplied Florida's indigenous people with a variety of beneficial products from food to fibers. The Seminoles still craft their well-known dolls from saw palmetto fibers. Saw palmetto is one of the leading U.S. herbal medicines due to the fruit's efficacy in treating benign prostate swelling. Several important questions remain unanswered. Better estimates of annual yield are needed as well as detailed economic analyses of this non-timber forest product. Researchers have yet to estimate sustainable harvest levels of palmetto fruits and the effects of harvest on wildlife. Geographical and temporal variations in fruit production and liposterol content are unknown, as are the fruit's active constituents. Our ongoing research addresses some of these questions.

ACKNOWLEDGMENTS

We thank Florida International University and the Seminole Tribe of Florida for financial support of this research. Billie Cypress and David Blackard facilitated work in the Big Cypress Seminole Reservation. Agness Cypress, Theresa Cypress, Sonnie Billie, Mary Frances Johns, and Alice Snow shared their knowledge of the Seminole use of saw palmetto. Marlin Huffman, president and C.E.O. of Plantation Medicinals, kindly provided a wealth of information on palmetto fruit industry in Florida. We thank Dr D. F. Austin for providing an image of saw palmetto fruits. Bill Popin and 2 anonymous reviewers provided useful comments on drafts of the manuscript. We dedicate this research to the late Leonard McCulloch who made the initial research among the Seminoles possible. He was an outstanding student of southern Florida's natural and cultural history and a good friend. Thanks, Leonard.

LITERATURE CITED

Abrahamson, W. E., and C. R. Abrahamson. 1989. Nutritional quality of animal dispersed fruits in Florida sandridge habitats. Bulletin Torrey Botanical Club 116:215–228.

Abrahamson, W. E., and D. C. Hartnett. 1990. Pine flatwoods and dry prairies. Pages 103–149 *in* R. L. Myers and J. J. Ewel, eds., Ecosystems of Florida. University of Central Florida Press, Orlando.

Anonymous. 1947. New uses for scrub palmetto. The Forest Farmer 6:6.

Bailey, L. H. 1945. Revision of the palmettos. Gentes Herbarium 6:1–459.

Benshoff, P. 1995. Palmetto poaching. Resource Management Notes 6:3.

Braeckman, J. 1994. The extract of *Serenoa repens* in the treatment of benign prostatic hyperplasia: A multicenter open study. Current Therapeutic Research 55:776–785.

Brevoort, P. 1995. The U.S. Botanical Market—An overview. HerbalGram 36:49–57.

Bruhwyler, J. 1994. Prostaserene—Purified extract of *Serenoa repens.* Drugs of the Future 19:452–453.

Bush, C. S., and J. F. Morton. 1963. Native trees and plants for Florida Landscaping. Dept. Agriculture–State of Florida, Bulletin No. 193.

Craig, R. M. 1991. Plants for coastal dunes of the Gulf Coast and South Atlantic Coasts and Puerto Rico. U.S.D.A. Soil Conservation Service, Agriculture Information Bulletin 460:1–41.

Craighead, F. C. 1971. Trees of South Florida, Vol. 1. The natural environments and their succession. University of Miami Press, Coral Gables, FL.

Cunningham, A. B., and F. T. Mbenkum. 1993. Sustainability of harvesting *Prunus africana* bark in Cameroon: A medicinal plant in international trade. People and Plants Working Paper 2. UNESCO, Paris.

Dickinson, J. 1945. Jonathan Dickinson's Journal or God's protecting providence being the narrative of a journey from Port Royal in Jamaica to Philadelphia, August 23, 1696 to April 1, 1697, edited by E. W. Andrews and C. Mc. Andrews. Florida Classics Library, Stuart, Florida.

Downs, D. 1995. Art of the Florida Seminoles and Miccosukees. University Press of Florida, Gainesville, FL.

Duke, J. A. 1985. CRC handbook of medicinal herbs. CRC Press, Boca Raton, FL.

Elias, T. S. 1980. The complete trees of North America: Field guide and natural history. Times Mirror Magazines, Inc., New York.

Farnsworth, N. R., and D. D. Soejarto. 1985. Potential consequences of plant extinction in the United States on the current and future availability of prescription drugs. Economic Botany 39:231–240.

Feurt, S. D. 1953. A Pharmacological study of *Serenoa repens* (Bartram) Small berries. Ph.D. dissertation, University of Florida, College of Pharmacy, Gainesville, FL.

Fisher, J. B., and P. B. Tomlinson. 1973. Branch and inflorescence production in saw palmetto (*Serenoa repens*). Principes 17:10–19.

Glassman, S. F. 1972. A revision of B. E. Dahlgren's Index of American Palms. Verlag von J. Cramer, Germany.

Grieve, M. 1971. A modern herbal—The medicinal, culinary, cosmetic, and economic properties, cultivation and folk-lore of herbs, grasses, fungi, shrubs and trees, with all their modern scientific uses. Dover, New York.

Hale, E. M. 1898. Saw palmetto (*Sabal serrulata, Serenoa serrulata*): Its history, botany, chemistry, pharmacology, provings, clinical experience and therapeutic applications. Boericke and Taffel, Philadelphia.

Hartwell, J. L. 1970. Plants used against cancer. A survey. Lloydia 33:288–318.

Hilmon, J. B. 1968. Autecology of saw palmetto, *Serenoa repens* (Bartr.) Small. Ph.D. dissertation, Duke University, Durham, NC.

Johnson, C. H. 1960. Important medicinal plants of Florida. State of Florida, Dept. Agriculture Bulletin No. 14.

Johnson, A. F., and M. G. Barbour. 1990. Dunes and maritime forests. Pages 429–480 *in* R. L. Myers and J. J. Ewel, eds., Ecosystems of Florida. University of Central Florida Press, Orlando.

Kushlan, J. E. 1990. Freshwater swamps. Pages 324–363 in R. L. Myers and J. J. Ewel, eds., Ecosystems of Florida. University of Central Florida Press, Orlando.

Long, R. W., and O. Lakela. 1976. A flora of tropical Florida. University of Miami Press, Coral Gables, FL.

Maehr, D. S., and J. N. Layne. 1996. Florida's all-purpose plant the saw palmetto. The Palmetto (Fall 1996):6–15, 22.

Maisch, J. M. 1899. A manual of organic materia medica, 7th edition. Lea Brothers and Co., Philadelphia.

McGoun, W. E. 1993. Prehistoric peoples of south

Florida. The University of Alabama Press, Tuscaloosa, AL.

Morton, J. F. 1977. Wild plants for survival in south Florida, 4th edition. Fairchild Tropical Garden, Coral Gables, FL.

———. 1986. Plants of long ago: Being mostly quotations from historical literature. Morton Collecteana Communication 4:1–35.

Murray, M. T. 1992. The healing power of herbs: The enlightened person's guide to the wonders of medicinal plants. Prima Publishing, Rocklin, CA.

Myers, R. L. 1990. Scrub and high pine. Pages 150–193 in R. L. Myers and J. J. Ewel, eds., Ecosystems of Florida. University of Central Florida Press, Orlando.

Neill, W. T. 1956. Florida's Seminole Indians. Great Outdoors Publishing Co., St. Petersburg, FL.

Pierce, C. W. 1970. Pioneer life in southeast Florida, edited by D. W. Curl. University of Miami Press, Coral Gables, FL.

Platt, W. J., and M. W. Schwartz. 1990. Temperate hardwood forests. Pages 194–229 in R. L. Myers and J. J. Ewel, eds., Ecosystems of Florida. University of Central Florida Press, Orlando, FL.

Radford, A. E., H. E. Ahles, and C. R. Bell. 1968. Manual of the vascular flora of the Carolinas. The University of North Carolina Press. Chapel Hill, NC.

Reitz, E. J., and C. M. Scarry. 1985. Reconstructing historical subsistence with an example from sixteenth-century Spanish Florida. Historical Archaeology, Special Publication Series 3:1–150.

Ritchason, J. 1994. The little herb encyclopedia: The handbook of nature's remedies for a healthier life. Woodland Health Books, Pleasant Grove, Utah.

Robinson, R. B. 1947. Minor fiber industries. Economic Botany 1:4–56.

Sargent, C. S. 1899. *Serenoa serrulata* as a remedy. Kew Bulletin of Miscellaneous Information 147–148:55–56.

Simpson, C. T. 1926. Ornamental gardening in Florida. Published by the author, Little River, Florida.

Small, J. K. 1926. The saw palmetto—*Serenoa repens.* Journal of the New York Botanical Garden 27:193–202.

Smith, D. 1972. Palm briefs: Fruiting in saw palmetto. Principes 16:30–33.

Sollmann, T. 1917. A manual of pharmacology and its applications to therapeutics and toxicology, 3rd Edition. W. B. Saunders, Philadelphia.

Sturtevant, W. C. 1955. The Mikasuki Seminole: Medical beliefs and practices. Ph.D. dissertation, Yale University, New Haven, CT.

Swanton, J. R. 1922. The early history of the Creek Indians and their neighbors. Bureau of American Ethnology, Bulletin 73:384–392.

Tasker, G. 1994. Enchanted ground: Gardening with nature in the subtropics. Andrews and McMeel, Kansas City, MO.

Tyler, V. E. 1982. The honest herbal: A sensible guide to the use of herbs and related remedies. Pharmaceutical Products Press, New York.

———. 1994. Herbs of choice: The therapeutic use of phytochemicals. Pharmaceutical Products Press, New York.

Uhl, N. W., and J. Dransfield. 1987. Genera Palmarum: A classification of palms based on the work of Harold E. Moore, Jr. The L. H. Bailey Hortorium and the International Palm Society, Allen Press, Lawrence, KS.

Widmer, R. J. 1988. The evolution of the Calusa: A nonagricultural chiefdom on the southwest Florida coast. The University of Alabama Press, Tuscaloosa, AL.

Wilder, E. A., and E. D. Kitzke. 1954. Waxy saw palmetto—A chemurgic opportunity. Chemurgic Digest 13:18–19.

Zona, S. 1997. The genera of Palmae (Arecaceae) in the southeastern United States. Harvard Papers in Botany 11:89–125.

TRADITION, USES AND BIODIVERSITY OF ROCKET (*ERUCA SATIVA*, BRASSICACEAE) IN ISRAEL[1,2]

ZOHARA YANIV, D. SCHAFFERMAN, AND Z. AMAR

Yaniv, Zohara; Schafferman, D. *(Dept. of Genetic Resources and Seed Research, A.R.O., The Volcani Center, Bet Dagan, Israel)* **and Amar, Z.** *(Dept. of Eretz Israel Studies, Bar-Ilan University, Ramat-Gan, Israel).* TRADITION, USES AND BIODIVERISTY OF ROCKET (*ERUCA SATIVA*, BRASSICACEAE) IN ISRAEL. *Economic Botany 52(4):394–400. 1998.* Eruca sativa *(rocket) is a native plant of Israel, documented in the old literature. A survey was conducted of the old literature of ancient Israel, including Jewish, Classical and Islamic sources up to the Middle ages. It was found that rocket was used as a garden crop and spice. It was also known as a medicinal plant and was used as an aphrodisiac, for eye infections, and for digestive and kidney problems. It is important to note that special attention was given in the old times to biodiversity within the rocket species. In the light of its high erucic acid content, native accessions were collected and evaluated as a potential future source of industrial oil. Ten accessions of* Eruca sativa *were cultivated in Bet-Dagan experimental farm during the 1995/96 growing season. Physiological as well as chemical parameters were recorded. Erucic acid (C 22:1) and eicosenoic acid (C 20:1) contents varied from 33 to 45% and from 7.3% to 9.8%, respectively. Special attention was given to the expressed biodiversity among the accessions and its relation to their origins.*

TRADITION, USAGES ET BIODIVERSITÉ DE ROQUETTE (*ERUCA SATIVA*, BRASSICACEAE) EN ISRAEL. Eruca sativa *(Roquette) est une plante endemique (native) en Israel, documentée dans la littérature ancienne. Une recherche de la littérature ancienne d'Israel comprenant des source juives, classiques et islamiques a été effectuée. Il s'est avéré que la roquette était utilisée comme plante de jardin et de condiment. Elle a été connue comme plante médicinale et utilisée comme aphrodisiaque, contre les infections des yeux et pour des problèmes digestifs et rénaux. Il est important de noter l'attention spéciale accordée dans les temps anciens à la biodiversité parmi les espéces de roquette. Tenant compte de sa haute contenance en acide erucique, des accessions natives ont été collectées et évaluées comme future source potentielles d'huile industrielle. Dix accessions* d'Eruca sativa *ont été plantées α la ferme expérimentale à Bet Dagan durant la saison de culture 1995/96. Des paramètres physiologiques et chimiques ont été enregistré. Les teneurs en acide erucique (C22:1) et en acide eicosenoique (C20:1) ont varié de 33 à 45% et de 7.3 à 9.8% respectivement. Une attention speciale a été acordée a la biodiversité exprimée dans les accessions et sa relation a ses origines.*

Key Words: Biodiversity; ethnobotany; rocket; seed-oils.

The name rocket (*Eruca sativa*) is well documented in the old literature of the Holy Land. It is called "Euzomon" in Greek, meaning good broth, for its flavoring qualities. In old Hebrew is called "Gargir" from which the Arabic "gargir" is derived. Most scientists agree that this is the garden vegetable mentioned in the Bible (King II 4:39–40) as "Oroth" (Feliks1976; Löw 1924: 491; Zohary 1982: 101).

[1] Received 22 March 1997; accepted 4 May 1998

[2] Contribution from the Agricultural Research Organization, The Volcani Center, Bet-Dagan, No. 2066-E, 1997 series.

Many sources are mentioned in the Mishna and the Talmud (a Jewish manuscript from the first-fifth centuries) about rocket's uses in the Holy Land during the Hellenistic period, as a spice, a food and a medicine. Josephus, the historian of the first century mentioned the plant in his description of the hat of the high priest (Josephus, Ant. III. 174, see Thackeray 1961).

The latest botanical monographers (Tutin 1964) and horticulturists (Akeroyd 1995) consider Rocket as *Eruca vesicaria* (L.) Cavanilles subsp. *sativa* (Mill.) Thellung with *E. sativa* Mill. in synonymy.

As part of an extensive effort to preserve the

Economic Botany 52(4) pp. 394–400. 1998

species of *Eruca sativa* in Israel a two-aspect research study was conducted: 1. the old literature of the Holy Land was searched in order to find evidence for the use and cultivation of Rocket during the Middle Ages; 2. native accessions of *Eruca sativa* were collected and their biodiversity was studied by comparing some chemical and physiological properties with those of some European lines.

Materials and Methods

Seed Collection

Seeds of *Eruca sativa* were collected from the wild, as part of a large on-going project of Crucifer collection in Israel.

Seeds were collected by two different expeditions led by two leading botanists: "Mattatia"—seeds were collected from two sites in the Jordan Valley representing the Saharo-Arabian region (based on the phytogeographical map of Israel (Plitmann et al. 1983: 15; Zohari 1996); b. "Yair"—seeds were collected from four sites in the Mediterranean region. At least 3 plants were collected from each site and seeds were kept separately for further research (Elber et al. 1989).

Oil quantity and fatty acid composition of the seeds were analyzed using the methods described by Yaniv et al. (1991).

Seed Introduction

Four seed samples of *Eruca sativa* from different botanical gardens were received in our genebank and used in the cultivation tests: Line No. 1/96 from Berlin Botanical Gardens; Germany; Line Nos. 2/96 and 6/96 from Turin Botanical Garden, Italy; Line No. 16/95 from Bari, Italy.

Cultivation

Ten lines (four introductions and six native accessions) of *Eruca sativa* were cultivated at the Bet Dagan Experimental Station. Each accession was replicated four times. Seeds of each accession were sown on Dec. 3. 1995 in 2.4 m2 plots consisting of four rows, with 30 cm between rows. Basic fertilization was done at the time of soil preparation, at a rate of 100:100:50 N:P:K. Treflan (2.5 kg/ha) was used as a herbicide. Irrigation was applied until seedling establishment. Date of emergence was Dec. 14. 1995.

Plants were harvested at seed maturity during the month of May 1996 (between 8 and 26 May), circa 150 days after emergence. Oil quantity and fatty acid analysis were conducted as described by Yaniv et al. 1991.

Lipid Extraction

Seeds were dried overnight at 50°C and ground into powder in a Moulinex coffee grinder. Five grams of powder were mixed with 100 cc petroleum ether (40°–60°C), and the lipid fraction was extracted in a Soxhlet apparatus for 16 h at 60°C. The solvent was evaporated, and the lipid fraction residues were weighed.

Direct Transesterification from Seeds

Seeds (200mg) were dried overnight at 50°C and ground into powder with mortar and pestle, after which 0.3 cm3 of dichloromethane and 2.0 cm3 of 0.5 N sodium methoxide (MeONa) were added. The tube was shaken and heated for 30 min at 50°C. The reaction was stopped by adding 5.0 cm3 of water containing 0.1 cm3 of glacial acetic acid.The esterified fatty acids were extracted with 2.0 cm3 petroleum ether (40°–60°C). The clear fraction was kept at −20°C until further analysis. Samples of 2.0 mm3 were injected into the gas-chromatograph for fatty acid analysis.

Gas Chromatography of Methylated Fatty Acids

A Megabore column (DB-23, 0.5 (m film thickness, 30 m × 0.54 mm, J&W Scientific) was used in a gas-chromatograph equipped with a flame ionization detector (Varian, model 3700 GC) and an automatic area integrator (3390A HP). The flow rate of N2 was 30 cc/min and the oven temperature was raised from 135 to 200(C, programmed at a rate of 4(C/min. The following fatty acids were identified by comparison with known standards (Supelco): C16:0, palmitic; C18:0, stearic; C18:1, oleic; C18:2, linoleic; C18:3, linolenic; C20:1, eicosenoic; and C22:1, erucic acid.

Results and Discussion

Traditional Uses

Table 1 summarizes the various uses of rocket in ancient Israel from Biblical times to the Middle ages. The relevant references are mentioned in Table 1 and cited fully in the bibliography. Classical sources cited for comparison are indicated in Table 1 with an asterisk.

TABLE 1. SOURCES FOR APPLICATIONS AND USES OF *ERUCA SATIVA* (ROCKET) FROM BIBLICAL TIMES TO THE MIDDLE AGES.

Period (Century)	Application							
	Eye infections	Aphrodisiac	Salad green & spices	Biological control	Deodorant & cosmetic uses	Digestion & kidney functions	Dog bites	Biodiversity
Biblical	Kings II							
I–III	Pliny*	Dioscorides* Galen*	Dioscorides* Mishnah, (Ma'asrot 4,8) Tosephta, (Shavit 2,9)	Mishnah, (Sheviit 9,1)		Dioscorides*		Mishnah, (Ma'asrot 4,8 Sheviit 4,8) Tosephta (Sheviit 2,9) Dioscorides*
III–V	Rabbi Yohana R. Huna (Talmud bavli) (Yoma 18b)							
V–VIII	B. Yoma 18b	B. Yoma 18b	B. Aroven 28b					B. Yoma 18b
VIII–XII		Harkavy			al-Tamimi			Meyerhof & Sabhy Ibn al-Awwam
XII–XVI		al-Qazwine	Tanhum Hayerushalmi Asali	al-Qazwine	Maimonides Ibn Wahsiyya Ibn al-Baytar		Dawud	Ibn Wahsiyya

Rocket in the Treatment of Eye Infection

Oroth, the biblical name for rocket means light. Rabbi Yohanan tells us that Oroth clears the eyes (Talmud, third century CE). Rabbi Huna recommands gargir for eye treatment (Talmud, third century CE). Evidence from Pliny in Rome (Jones1887: 49) supports the same use in southern Europe.

Rocket as an Aphrodisiac

Rocket was known as an aphrodisiac throughout ancient times. Both Dioscorides and Galen (Gunther 1959; Kühn 1826: 72) recommanded eating seeds for increasing semen production. Galen warned gainst its sharp taste and a possible headache and recommended eating seeds mixed with other foods.

In the Jewish literature we find the same indications for increasing semen production: in the Talmud (B. Yoma 18b) and in the Rabbinical literature of the tenth century of Irak (Harkavy 1887: 215). al-Qazwine of the thirteenth century indicated that eating seeds with honey will stimulate sexual desire. (al-Qazwine 1981: 244).

Rocket as a Salad-Green and a Spice

Rocket has been known since the first century CE as a salad-green and as a spice—the seeds were used as mustard seeds. According to Dioscorides (Gunther 1968: 181) the seeds were used immersed in milk. In the Jewish Mishnah of the same period it is mentioned that Rocket was used as a pepper substitute. The seeds were crushed and the paste was used to flavor meat . (Talmud, B. Aroven 28b). During the thirteenth century CE it was eaten in Israel as a vegetable named "gargir" and mentioned by Tanhum Hayrushalmi (meaning—from Jerusalem) (Toledano 1961: 84). In 1391, a sacred moslem document from Jerusalem was found with a list of a dead man's belongings, including a container called "Gergir". (al-Asali 1983: 34).

Rocket as a Deodorant and for Cosmetic Uses

Rocket is mentioned for cosmetic uses by al-Tamimi, a Jerusalem physician of the tenth century. He wrote that eating Rocket on an empty stomach prevents sweat smells (Ibn al-Baytar, 1875: 160). Maimonides and Ibn Wahsiyya recommanded using the ground seeds as a cream on the face for the treatment of acne. Eating the seeds or spreading their ground powder under the arms has a deodorant effect.

Rocket as a Digestion-Aid and Kidney Mulfunction

Mentioned by Dioscorides in the first century.

Rocket and Dog-Bites

Mentioned by Dawud (1935: 106) in the Middle Ages. A decoction is prepared for protecting against dog-bites.

Biological Control

In the literature of the thirteenth century (al-Qazwine 1981: 244) there is an indication of the practice of biological control—Rocket seeds were sown together with other garden vegetables in order to inhibit pest development.

Biodiversity in Tradition

It is interesting to note that in the ancient literature the concept of differentiating between native plants and cultivated plants, and the effect of the origin of the germplasm on the quality of the plants is quite developed. Dioscorides, first century (Gunther1968: 181) has a remark about the pungency of the wild type. In the Jewish reference of Mishna (first century, Sheviit 9,1,) the wild type is called "a wild seed" to differenciate it from the more important cultivated greens, and in the medical literature of the Middle Ages various accessions of Rocket are described (Meyerhof and Sobhy 1937) and instructions for cultivation are included. (Clement-Mullet 1865: 301; Fahd 1995)

Biodiversity in Field Experiments

Data regarding some physiological parameters of 10 lines of *Eruca sativa* cultivated at Bet-Dagan Experimental Station are summarized in Table 2. This table illustrates the biodiversity expressed within the lines. In spite of uniform sowing and emergence dates, (see Materials and Methods) there was a significant variability in the onset of flowering and of ripening. There were differences of 27 days between early and late flowering types, and of 18 days between early and late ripening. It is interesting to note that the "Yair" native accessions (3/96,20/96,21/96,22/96) were the earliest to flower and to ripen. The "Mattatia" collection lines were closer to the European lines, both in flowering and ripening dates and in seed color. Seed

TABLE 2. FLOWERING AND RIPENING TIMES OF 10 LINES OF *ERUCA SATIVA* CULTIVATED AT THE BET—DAGAN EXPERIMENTAL STATION (1995/96)

Line no.	Origin	No. of days from emergence to: Flowering	Ripening	Seed color	1000—seed weight (g)
1/96	Germany	*77 c**	163 b	light brown	1.55 d
2/96	Italy	73 bc	163 b	light brown	1.47 e
6/96	Italy	71 d	163 b	light brown	1.80 b
16/95	Italy	88 d	163 b	light brown	1.30 f
14/96	Israel-Matt.	74 bc	163 b	light brown	1.72 c
18/96	Israel-Matt.	70 b	163 b	light brown	1.69 c
3/96	Israel-Yair	66 ab	146 a	dark grown	1.45 e
20/96	Israel-Yair	66 ab	145 a	dark brown	1.72 c
21/96	Israel-Yair	62 a	145 a	dark brown	1.48 e
22/96	Israel-Yair	60 a	145 a	dark brown	1.88 a

* Nos. are averages of four replications.
** Different letters indicate significant differences for each parameter separately at $P < 0.05$ (Duncan's Multiple Range Test).

weight varied from 1.3 to 1.9g/1000 seeds, with no clear correlation with their origin.

Table 3 shows data regarding oil content and the fatty acid composition in seed oils of ten *Eruca sativa* lines: Oil content varies from 24.5 to 29.2%. The three top lines are European introductions. There is a genetic difference in oil content of seeds but this trait could be improved by selection and breeding.

Eruca sativa, as a member of the Brassicaceae, contains erucic acid (C22:1), a fatty acid unique to seed oils. The erucic acid content in the oil varies between 33% (in native Israeli lines) and 47% (in the introduced line from Germany) (Table 3). Lines with high erucic acid contents have lower level of linoleic acid (C18:2) and linolenic acid (C18:3) contents. These fluctuations are indications of biodiversity present in the species related to the the geographic origin of the collected accessions.

Table 4 represents the ratio of mono-unsaturated fatty acids (oleic, C18:1; eicosenoic C20:1 and erucic, C22:1) to polyunsaturated fatty acids (linoleic, C18:2; and linolenic, C18:3). It is interesting to note that the four Israeli accessions, collected by Yair (Mediterranean region) are very uniform in their fatty acid profiles (lines 3/96, 20/96, 21/96, 22/96). The European introduction are different from those, but are also uniform among themselves. The two native lines collected from the Jordan Valley are closer to the European group than to the other Israeli

TABLE 3. FATTY ACID COMPOSITION IN SEED OILS OF *ERUCA SATIVA* LINES (INTRODUCTIONS AND NATIVE FLORA)

Line no.	Origin	Oil content %	Fatty acid (% of total) Palmitic C 16:0	Stearic C 18:0	Oleic C 18:1	Linoleic C 18:2	Linolenic C 18:3	Eicosenoic C 20:1	Erucic C 22:1
1/96	Germany	29.1 a	*5.1 a**	1.3 b	15.1 a	8.3 b	14.7 b	7.4 b	44.7 a
2/96	Italy	28.7 a	4.9 a	1.4 b	16.7 a	10.3 b	14.6 b	7.3 b	42.8 a
6/96	Italy	27.8 ab	4.8 a	1.6 b	15.2 a	9.4 b	15.2 b	7.6 b	43.3 a
16/95	Italy	28.8 a	4.8 a	1.4 b	15.9 a	9.5 b	15.1 b	7.6 b	43.3 a
14/96	Israel-Matt.	27.8 ab	4.7 a	1.4 b	17.8 a	9.4 b	14.9 b	7.5 b	42.4 a
18/96	Israel-Matt.	27.8 ab	5.2 a	1.4 b	16.9 a	9.6 b	14.9 b	7.4 b	41.7 a
3/96	Israel-Yair	24.8 c	4.1 b	1.5 b	9.9 b	14.1 a	19.7 a	9.8 a	37.0 b
20/96	Israel-Yair	26.2 b	4.0 b	1.9 a	11.4 b	14.0 a	18.8 a	9.4 a	36.3 b
21/96	Israel-Yair	25.6 bc	4.2 b	1.8 a	11.5 b	14.8 a	19.2 a	9.6 a	35.2 b
22/96	Israel-Yair	25.5 bc	4.0 b	1.9 a	12.6 b	15.3 a	18.9 a	9.4 a	33.4 b

* Numbers are averages of four replications.
** Different letters indicate significant differences for each parameter separately at $P < 0.05$ (Duncan's Multiple Range Test).

TABLE 4. RATIO OF MONOUNSATURATED/POLYUNSATURATED FATTY ACID IN *ERUCA SATIVA* LINES.

Line no.	Origin	Fatty acid (% of total)	
		C 18:1 + C 20:1 + C 22:1	C 18:2 + C 18:3
1/96	Germany	67.2	23.0
2/96	Italy	66.8	24.9
6/96	Italy	66.1	24.6
16/95	Italy	66.8	24.6
14/96	Israel-Matt.	67.7	24.3
18/96	Israel-Matt.	66.0	24.5
3/96	Israel-Yair	56.7	33.8
20/96	Israel-Yair	57.1	32.8
21/96	Israel-Yair	56.3	34.0
22/96	Israel-Yair	55.4	34.2

lines. There is a reciprocal correlation between the levels of monosaturated and polyunsaturated fatty acids in the seed oils. In a study by our group involving the analysis of the diversity of native *Sinapis alba* accessions, a genetic variability among eight accessions collected from two geographical locations was demonstrated by RAPD (Random Amplified Polymorphic DNA) markers. (Granot et al. 1996). A genetic distance between accessions from the two locations was found. In addition, RAPD analysis revealed a genetic link between *Sinapis alba* genotypes and their erucic acid content. These results emphasis the importance of preserving and documenting the natural biodiversity in the species *Eruca sativa.*

LITERATURE CITED

Akeroyd, J. R. 1995. Eruca. Page 161 *in* J. Cullen, J. C. M. Alexander, A. Brady, C. D. Brickell, P. S. Green, V. H. Heywood, P. M. Jörgensen, S. L. Jury, S. G. Knees, A. C. Leslie, V. A. Matthews, N. K. R. Robson, S. M. Walters, D. O. Wijnands, and P. F. Yeo, eds. The European Garden Flora. Vol. IV. Cambridge Univ. Press, Cambridge.

al-'Asali, K. J. 1983. Wathaik Makdisiyyah tarikhyya, I/2, Jordan Academy Pub., Aman.

Dawud B., Umar al-Antaki, 1935. Tadkhirai Uli al-Albab, Cairo (no publisher is indicated)

Clement-Mullet, J. J., ed. 1865. Le livre de l'agriculture d'Ibn al-Awwam. II. Paris (no publisher is indicated).

Elber, Y., Z. Yaniv, D., Schafferman, E. Ben-Moshe and M. Zur. 1989. List of wild crucifer seed collection. The Israel gene bank for agricultural crops. (internal publication), Bet Dagan, Israel.

Fahd, T. ed. 1995. L'Agriculture Nabateenne of Ibn Wahsiyya, II. Institute Francais de Damas Pub. Damascus. pp. 779–780

Feliks, Y. 1976. Plant World of the Bible, Bar-Ilan University Press, Ramat-Gan. pp. 190–191

Granot, D., E. Shabelsky, D. Schafferman and Z. Yaniv. 1996. Analysis of genetic variability between populations of *Sinapis alba* and the effect of cultivation on the variability. Acta Horticulturae 407:67–74.

Gunther, R. T. 1968. The Greek herbal of Dioscorides—de materia medica. II/170, Hafner Publishing Co., New York.

Harkavy, A. 1887. Teshuvot ha-geonim, Tzvi Hirch Tskavsky Pub., Berlin.

Ibn al-Baytar.1875. Kitab al-Jami'li-mufradat al-adwiya wal-aghdiya, I. Bulaq Pub., Cairo.

Jones, W. H. S., trans. 1887. Historia naturalis of Pliny. XX. The Loeb Classical Library Pub., London-Cambridge.

Kühn, C. G. 1826. Claudii Galeni opera omnia, XII. Leipzig.

Löw, I. 1924. Die Flora der Juden, I, Vienna-Leipzig.

Meyerhof, M., ed. 1940. Un Glossaire de Matiere Medicale Compose par Maimonide,MIE Pub., Cairo. 74:40

Meyerhof, M. and Sobhy, G. P. 1937. The abridged version of "The Book of simple drugs" of Ahmad Ibn Muhammad al-Ghafiqi by Gregorius Abul-Farag (Barhebraeus), No. 202 pp. 409–411, al-Ettemad Printing Press, Cairo.

Plitmann, U., C. Heyn, A. Danin and A. Shmida. 1983. Pictorial flora of Israel. Massada Publishers, Israel.

al-Qazwine. 1981. Agaeib al-Mkhlukat, Beit-Hamizrah Haravi. Pub. Beirut.

Thackeray, H. J., trans. 1961. Jewish Antiquities of Josephus. III. The Loeb Classical Library, London. 7:174

Toledano, B. 1961. "al Mursid al Kafi" of Tanhum Hayrushalmi, Tel-Aviv.

Tutin, T. G. 1964. Eruca. Page 340 *in* T. G. Tutin, V. H. Heywood, N. A. Burges, D. H. Valentine, S. M.

Walters, and D. A. Webb, eds., Flora Europaea. Vol. I. Cambridge Univ. Press, Cambridge.

Yaniv, Z., Y. Elber, M. Zur and D. Schafferman. 1991. Differences in fatty acid composition of oils of wild Cruciferae seed. Phytochemistry 30:841–843.

Yaniv, Z., D. Schafferman, Y. Elber, E. Ben-Moshe and M. Zur. 1993. Evaluation of *Sinapis alba*, native to Israel, as a rich source of erucic acid in seed oil. Journal of Industrial Crops 2:137–142.

Zohary, M. 1966. Flora Palaestina. Part 1, Israel Academy of Sciences and Humanities, Jerusalem. pp. 246–329.

Zohary, M. 1982. Plants of the Bible. Cambridge University Press, London.

Book Review

Essential Oil Crops. E. A. Weiss. 1997. CAB International, 198 Madison Avenue, New York, NY 10016-4341. xii + 600 pp. (hardcover). U.S. $135.00. ISBN 0-85199-137-8.

The book is divided into 15 chapters. Chapter one is devoted to "World Trade in Essential Oils." This discussion provides information on the major producers (Asian Countries, Brazil, China, Egypt, India, Indonesia, USA), and major importers (European Union Members, Japan, Switzerland, USA) of aromatic materials. Also given are data on the volume of trade per product. Chapter 15, "Distilling and Extracting Essential Oils," discusses and diagrams the various methods of obtaining essential oils.

Chapters 2 through 14 provide full information (history, botany, ecology, soil and fertilizers, cultivation, harvesting, distillation, pests and diseases, products and specifications) on many plant species. The text also provides partial information on other species, e.g., *Cananga* (Annonaceae), *Geranium* and *Pelargonium* (Geraniaceae), *Cymbopogon* and *Vetiveria* (Gramineae), *Pogostemon* (Lamiaceae), *Cinnamomum, Ocotea, Laurus, Litsea* and *Sassafras* (Lauraceae), *Myristica* (Myristicaceae), *Eucalyptus, Melaleuca, Leptospermum, Pimenta* and *Syzygium* (Myrtaceae), *Jasminum* and *Osmanthus* (Oleaceae), *Piper* (Piperaceae), *Rosa* (Rosaceae), *Amyris, Boronia* and *Citrus* (Rutaceae), *Osyris* and *Santalum* (Santalaceae), and *Zingiber* (Zingiberaceae). Common names are also provided.

For unexplained reasons, the author included cloves in the genus *Syzygium*. Past and recent treatments of the Myrtaceae include cloves in *Eugenia*.

In his introductory chapter, Weiss indicated that "the crops selected are those that the author believes should be encouraged in countries where cash crops are limited, or outside inputs to raise rural incomes difficult to obtain. For this reason, the important pine oils have not been included since these are produced in more developed countries."

In our opinion, if the author wanted to help developing countries, he should have provided available information regardless of where the compounds are produced. We say this because the bulk of production of essential oil is in developing countries. His statement also gives an impression that, if oils are produced in developed countries, developing countries should not become involved. This idea could be challenged. Moreover, if the author wanted to promote the production of aromatic plants from the third world, he should have included several commonly grown aromatic plants from those regions. A few examples are *Carum, Coriandrum* (Apiaceae), *Artemisia* (Asteraceae), *Ocimum, Mentha, Satureja* (Lamiaceae), *Aframomum, Cardamomum* and *Cucurma* (Zingiberaceae).

Furthermore, because world consumption of flavors and fragrances is to exceed U.S. $8 billion by the year 2000, Weiss suggests that developed countries should deliberately support essential oil production in less developed countries. Such support, he claims, will reduce demand or aid and limit north-south confrontation. That idea would have been more informative if data were provided showing where the money goes. How much of the money from aromatic plants, for example, would go to the primary producers (the third world countries), and how much of it to those companies who are reprocessing (developed countries). We also believe that the north-south confrontation on trade will be minimized when companies from the third world are properly compensated for their produce.

Although several of the ideas in this book are debatable, the data on plants and their oils is valuable and well documented. We recommend the book for professionals and amateurs interested in *Essential Oil Crops*.

Sebsebe Demissew and Nigist Asfaw
The National Herbarium, Science Faculty
Addis Abeba University
P.O. Box 3434
Addis Abeba, Ethiopia

VEGETATIVE, REPRODUCTIVE, AND PHYSIOLOGICAL ADAPTATIONS TO ARIDITY OF PITAYO (*STENOCEREUS QUERETAROENSIS*, CACTACEAE)[1]

EULOGIO PIMIENTA-BARRIOS AND PARK S. NOBEL

Pimienta-Barrios, Eulogio *(Departamento de Ecología, CUCBA, Universidad de Guadalajara, Jalisco, Mexico)* **and Park S. Nobel** *(Department of Biology, University of California, Los Angeles, California 90095-1606).* VEGETATIVE, REPRODUCTIVE, AND PHYSIOLOGICAL ADAPTATIONS TO ARIDITY OF PITAYO (*STENOCEREUS QUERETAROENSIS*, CACTACEAE). *Economic Botany 52(4):401–411. 1998. Pitayos (*Stenocereus spp.*) are columnar cacti that produce attractive edible fruits under both wild and cultivated conditions. Until recently the study of pitayos has been relatively neglected even though they have been a crucial staple for the inhabitants of subtropical semi-arid lands of Mexico since ancient times and this century have become an important fruit crop there. The main objective of this review is to relate seasonal aspects of vegetative and reproductive growth, CO_2 uptake, and carbohydrate resources to abiotic components of the environment. Flower and fruit production occur during the dry season in the late winter and spring, roots grow during the summer wet season, and stem elongation commences in the autumn at the beginning of the dry season. Vegetative growth thus does not coincide with reproductive growth, as it does for other fruit crops in temperate and tropical regions. The time offset of vegetative and reproductive activities reduces competitive sink effects, regulating assimilate partitioning to different organs, which may allow* Stenocereus queretaroensis *to adapt to water-limited environments. Reserve carbohydrates (starch, mucilage) accumulate in the early summer just after reproductive growth and in the early winter just after stem extension. Reducing sugars increase during the middle of the summer before stem elongation begins. The low rates of growth and of photosynthesis for* S. queretaroensis *are associated with low tissue levels of nitrogen, chlorophyll, and some micronutrients (Fe, Mn). Low levels of gibberellic acid may also contribute to its low growth rate. It does not respond morphologically to irrigation during the dry period, indicating a low plasticity for growth. Although* S. queretaroensis *is cultivated, its physiological and phenological activities closely resemble those of wild CAM species and other wild perennial plants, reflecting its rather recent domestication and the fact that selections were based mainly on fruit quality and adaptation to aridity rather than biomass productivity.*

ADAPTACIONES VEGETATIVAS, REPRODUCTIVAS Y FISIOLOGICAS DEL PITAYO (STENOCEREUS QUERETAROENSIS, CACTACEAE) A LA ARIDEZ. *Los pitayos (*Stenocereus spp.*) son cactáceas columnares que producen frutos comestibles de colores atractivos en poblaciones silvestres y cultivadas. Hasta hace poco tiempo el pitayo había sido poco estudiado, no obstante haber sido un alimento importante para los habitantes de las tierras semiáridas subtropicales desde tiempos antiguos y haberse convertido en este siglo en una importante especie frutal. El principal objetivo de esta revisión es relacionar el crecimiento vegetativo y reproductivo, la asimilación de CO_2 y las reservas de carbohidratos con algunos componentes abióticos del ambiente. La producción de flores y frutos ocurre durante la estacion seca del año. Las raíces crecen durante el periodo de lluvias en el verano y la elongación de las ramas comienza en el otoño al inicio de la estación seca. El crecimiento vegetativo no coincide con el crecimiento reproductivo, como ocurre con otras especies frutales de regiones templadas y tropicales. La competencia entre demandas metabólicas se reduce, debido a que no se traslapa el crecimiento vegetativo con el reproductivo, de esta manera se regula la distribución de fotosintetizados a los diferentes órganos de la planta, lo cual permite que* S. queretaroensis *se adapte en ambientes en los que el agua es un factor limitante. Los carbohidratos de reserva (almidón, mucílago) se acumulan al empezar el verano después del crecimiento reproductivo y en el invierno, justo después de la elongación de las ramas. El contenido de azúcares reductores aumenta a mediados del verano, antes del crecimento de las ramas. Las bajas tasas de crecimiento y fotosíntesis registrados en* S. queretaroensis *se asocian con niveles bajos de nitrógeno, clorofila y algunos micronutrientes como Fe y Mn. Probablemente niveles bajos de ácido giberélico también con-*

[1] Received 28 October 1997; accepted 6 May 1998

Economic Botany 52(4) pp. 401–411. 1998

tribuyan a la baja tasa de crecimiento de esta planta. S. queretaroensis *no responde morfológicamente al riego durante la estación seca, revelando poca plasticidad en el desarrollo. Aunque* S. queretaroensis *se encuentra en condición cultivada, su comportamiento fenológico y fisiológico presenta semejanza con especies silvestres con metabolismo CAM y otras plantas silvestres perennes. Este comportamiento refleja su relativamente reciente domesticación, así como el hecho de que al hacer la selección de fenotipos sobresalientes de pitayo se prestó una mayor atencion a la calidad del fruto y la adaptación a la aridez, en lugar de la productividad de biomasa.*

Key Words: Adaptation; aridity; cacti; carbohydrates; fruit.

The effects of aridity on food production has become a major concern in developing countries, especially in Latin America, because arid and semi-arid regions occupy one third of their area and most of the land available for the expansion of agricultural activities is in or bordering such regions (Izquierdo and Oltremari 1996). One factor limiting the development of agriculture in arid regions is the high sensitivity of most domesticated crops to drought. Recently, interest has increased for cacti that produce edible fruits, especially because such plants are resistant to drought. Important fruits are cactus pear (*Opuntia* spp.), which is presently cultivated in nearly 20 countries, and pitaya (*Stenocereus* spp.) (Mizrahi, Nerd, and Nobel 1997). Genetic resources for such potential fruit crops occur in subsistence agriculture in arid and semi-arid regions of Mexico and other countries.

This article reviews existing and new data on biological and agronomic aspects of *Stenocereus queretaroensis,* a columnar cactus that produces attractively colored edible fruits under both wild and cultivated conditions. *S. queretaroensis* is native to the subtropical semiarid lands of Mexico and it is widely distributed in the central part of Mexico (Bravo 1978: 590–591). Since prehispanic times, *Stenocereus* have provided valuable additions to the diets of Indian tribes inhabiting tropical and subtropical semiarid lands of Mexico. Recently, pitayo has emerged as a fruit crop whose cultivation is feasible using relatively low inputs of anthropogenic energy or water and whose fruits ripen before local markets are supplied with other summer fruits (Pimienta-Barrios and Nobel 1994). These agronomic traits have increased the economic viability of small farms in semi-arid subtropical lands of Mexico.

Phenology

Shoot growth for *S. queretaroensis* takes place at the branch apices and occurs essentially only in the autumn and early winter (October through December; Fig. 1a). Flower differentiation starts in early February, ending in early April; fruits ripen from early March until the end of May. Thus vegetative and reproductive growth occur in the dry season (Fig. 1b,c; Lomeli and Pimienta 1993; Pimienta-Barrios and

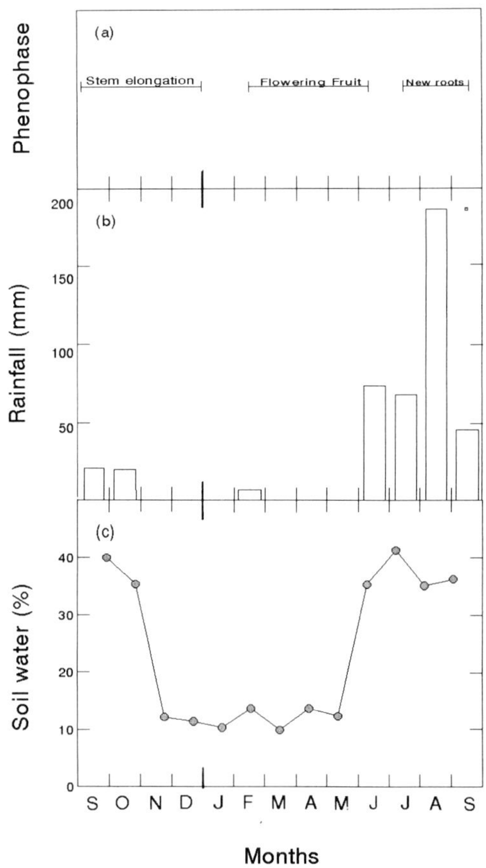

Fig. 1. Occurrence of the main reproductive and vegetative phenophases of *Stenocereus queretaroensis* (a), total monthly rainfall (b), and monthly soil moisture content (c) during 1994–1995 in the Sayula Basin, Jalisco, Mexico. Adapted from Pimienta, et al. (1998).

Nobel 1995). The differentiation of new roots coincides with the start of the summer rainy season, which is the only prominent vegetative event then (Fig. 1a; Arceta 1997). Such phenological behavior contrasts with that of woody perennials growing in temperate climates, whose root growth begins in the early spring when soil temperatures are rising and precedes stem growth, which occurs at the beginning of the spring and during the summer (Kozlowski, Kramer, and Pallardy 1991: 13–15). Moreover, vegetative growth of *S. queretaroensis* does not coincide with reproductive growth, as is does for most conventional crops in temperate and tropical regions (Ryugo 1988: 68–82). The time offset of vegetative and reproductive activities reduces competitive sink effects of different organs, which may help *S. queretaroensis* adapt to water-limited environments (Pimienta-Barrios and Nobel 1995).

Although *S. queretaroensis* is native to subtropical climates, water availability as well as air temperatures may not strongly influence its vegetative growth, as occurs for the majority of perennial species (Larcher 1995: 301–309), because the highest rates of primary growth occur during the middle of the autumn when both soil moisture and temperature can become limiting. Moreover, greater rates of stem extension were recorded in 1995 than in 1994 even though water availability was greater during the autumn of 1994 than of 1995 for *S. queretaroensis* in natural populations (Arceta 1997). Irrigation during the period of reproductive development of *S. queretaroensis* affects some reproductive traits. For instance, it increases flower and fruit size but decreases the number of flowers per plant by 36%. Irrigation decreases the number of ovules by 12%, seed number per fruit by 25%, and final seed set by 15%, although seed size is unaffected (Pimienta-Barrios, Robles-Murguia, and Dominguez de la Torre 1995).

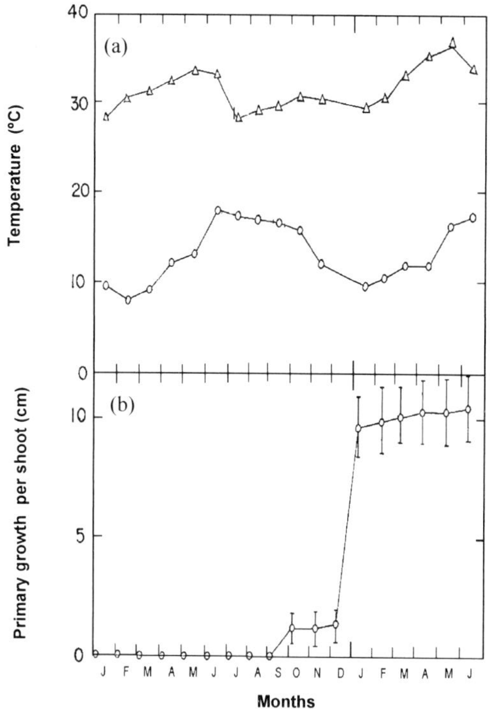

Fig. 2. Daily minimum (O) and maximum (D) air temperatures averaged over a month (a) and cumulative monthly stem extension for *S. queretaroensis* (b) in the Sayula Basin, Jalisco, Mexico in 1993–1994. Extension data are means ± SE, except when the error bars are smaller than the symbols (n = 24 stems). Adapted from Pimienta-Barrios and Nobel (1995).

Vegetative and Reproductive Growth

Shoot elongation in temperate perennial species, especially trees, is more rapid during the long days of late spring and early summer than during the autumn or winter (Salisbury and Ross 1992: 341–342). However, the major stem extension for *S. queretaroensis* occurs in the autumn when minimum air temperatures are lower and soil water is less available (Figs. 1, 2). For mature cultivated plants, 85% of the annual growth is from December to January, when the daily rate averages 0.26 cm day^{-1} (Fig. 2; Pimienta-Barrios and Nobel 1995). Thus *S. queretaroensis* is a slowly growing plant similar to other columnar cacti producing edible fruits (Nerd, Raveh, and Mizrahi 1993), but unlike young cladodes of *Opuntia* spp. under similar conditions (Nobel 1988: 184–186; Robles 1994) or other herbaceous and woody plants (Grime and Hunt 1975), which have higher growth rates. It thus behaves as do wild perennial plants growing in infertile soils that commonly tend to be long-lived with slow growth and low capacity for photosynthesis and nutrient absorption (Chapin 1980). Such plants often maintain growth and reproduction even during periods of low water availability (Grime 1979: 45–50) but tend to be less plastic than plants with higher growth rates (Grime and Hunt 1975). Low plasticity is indeed a feature of cultivated *S. quere-*

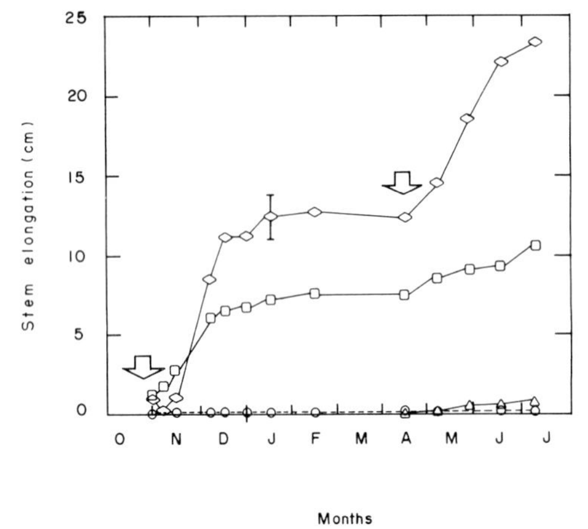

Fig. 3. Cumulative monthly stem extension for *S. queretaroensis* approximately 100 years old injected with 0 (○), 10^{-5} (△), 10^{-4} (□), or 10^{-3} (◇) M GA_3 on the two dates indicated by arrows. Data from 1994–1995 are means ± SE (n = 7 plants). Adapted from Pimienta, et al. (1998).

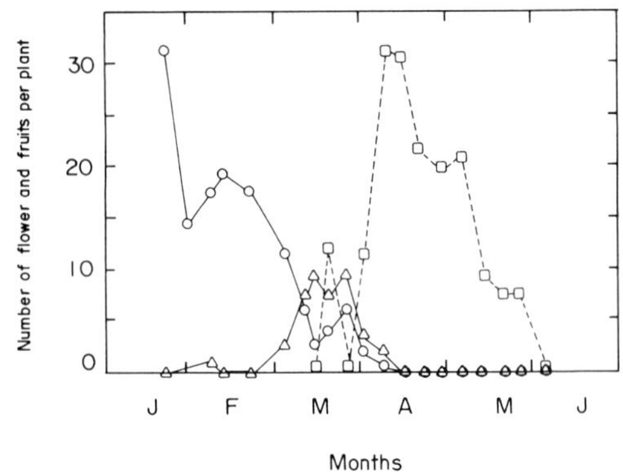

Fig. 4. Reproductive development for *S. queretaroensis* under natural conditions. Number of emerging flower buds (○), open flowers (△), and nearly ripe fruits (□). The decline in number toward the end of the season was the result of periodic fruit harvesting. Data from 1994 are means ± SE (n = 6 to 7 plants). Adapted from Pimienta-Barrios and Nobel (1995).

taroensis, because irrigation does not markedly affect stem extension, even when other environmental factors such as temperature and irradiance are favorable for photosynthesis and growth (Pimienta-Barrios and Nobel 1995). The slow growth rates of *S. queretaroensis* have limited its domestication, as 8 to 10 years are required to obtain a large enough production for commercial profitability, even when stem cuttings are used as the starting material for propagation (Pimienta-Barrios and Nobel 1994).

The low growth rates for cultivated *S. queretaroensis* are associated with low tissue levels of nitrogen, chlorophyll, and some micronutrients (Fe, Mn), even in deep alluvial soils (Nobel and Pimienta-Barrios 1995). In addition, low levels of hormones such as gibberellic acid may also contribute to the low growth rates. For instance, injecting gibberellic acid into stems of 100-year-old plants that have stopped growing, causes stem elongation to resume within 1 month (Pimienta-Barrios, et al. 1998; Fig. 3).

The rate of flower extension growth for *S. queretaroensis,* which is pollinated primarily by bats, is 0.31 cm day^{-1}, and 22 days elapse from the emergence of flower buds to anthesis. After abscission of the perianth, the developing fruit is about 15% longer for irrigated plants compared with those under natural rainfall conditions (Pimienta-Barrios, Robles-Murguia, and Dominguez de la Torre 1995). Fruit length attains over 90% of its final size in about 40 days, and the overall development period from anthesis to maturity is about 90 days, resulting in larger flowers under irrigation. The length enhancements resulting from irrigation during the last stages of flower and fruit growth suggest that such growth occurs by cell elongation, as for other fleshy fruits (Coombe 1976; Zanchin, et al. 1994). The time to reach maturity for fruits of *S. queretaroensis* is relatively short compared with other cacti (e.g., *Opuntia* spp.) and other fruit crops (peach, plum, apple, and avocado) (Barbera, Carimi, and Inglese 1992; Lee and Young 1983; Ryugo 1988: 107–111).

Reproductive Biology

Reproductive development of *S. queretaroensis* is asynchronous, as flowers in early stages of differentiation, flowers in anthesis, and young developing fruits occur simultaneously (Fig. 4). Asynchronous growth, which may be advantageous ecologically as well as commercially, is not affected by irrigation. The time between flower bud initiation and anthesis is relatively short for *S. queretaroensis,* about 40–50 days, as in other columnar cacti that produce edible fruits (Nerd and Mizrahi 1997; Pimienta-Barrios and Nobel 1994). This behavior is similar to that for other tropical and subtropical fruit crops, such as orange, avocado, mango, and cactus pear (Pimienta 1990; Osuna and Pimienta 1992), but it contrasts with that of temperate fruit crops (e.g., apple, peach and pear), whose flower bud initiation usually occurs 1 year before flowering (Faust 1989).

The occurrence of fruit ripening at the end of the spring may be optimal for *S. queretaroensis.* In particular, seed maturation occurs just before the start of the summer rainy season, which favors seed germination and establishment because of favorable soil moisture and shading by surrounding natural vegetation. As for other cacti (Nobel 1988: 87, 135–137; Nobel 1994: 104–106), nurse plants are necessary for the survival of *Stenocereus* seedlings. In contrast, seed development for *Opuntia* spp. in the same regions occurs near the middle or the end of the rainy season, so its seed maturation and dispersal coincide with the beginning of the dry season (Perez 1993). To avoid drought stress, the seeds of *Opuntia* spp. have a dormancy period of 7 to 8 months; its seed germination is initiated at the start of the next rainy season (Perez 1993).

The seeds of *S. queretaroensis* are smaller (2.6 mg) than those of *Opuntia ficus-indica* (23 mg), which is advantageous from a commercial point of view. They contain a relatively large amount of lipid and germinate in response to light. Seeds of many wild species that respond to light are also small and rich in lipids. Because of their small size and light requirements, such seeds must be near the soil surface for successful germination (Salisbury and Ross 1992: 451), whereas seeds of *O. ficus-indica* are not light-requiring and can germinate even when buried in the soil. The germination rate is higher for *S. queretaroensis* receiving irrigation). Seeds of S. queretaroensis are thus similar to those of wild species (Salisbury and Ross 1992: 451), reflecting its relatively recent domestication for its fruits.

Stem Anatomy

For both *S. queretaroensis* and *Opuntia* spp., the stomatal frequency is low (10–40 per mm^2; Pimienta, Loera, and Lopez 1993), which is generally considered to be an adaptation to arid environments. Cuticular thickness of the stems varies from 13 to 17 μm for *S. queretaroensis* and from 10 to 50 μm for *Opuntia* spp. (Jimenez-Lopez, Pimienta-Barrios, and Muñoz-Urias 1995; Pimienta-Barrios, Robles-Murguia, and Dominguez de la Torre 1995). Thus the cuticle of *S. queretaroensis* is relatively thick compared with that of mesophytic plants (Nobel 1994: 92) but thinner than for various species of *Opuntia.* A thick cuticle represents an important adaptation to aridity, leading to the ability to withstand long droughts (Nobel 1994: 92). Calcium oxalate crystal (druses) are not present in the stem cortex of *S. queretaroensis* consistent with its low calcium content (Nobel and Pimienta-Barrios 1995) but in contrast to other cacti (Gibson and Nobel 1986: 52; Nobel 1983).

A distinctive feature of stem anatomy of *S. queretaroensis* is the presence of aerenchyma in its hypodermis. Aerenchyma, characterized by continuous gas spaces in shoots and roots, is often an adaptation to low oxygen levels and is a distinctive feature of many wetland species (Esau 1977: 259; Salisbury and Ross 1992: 285–286). The aerenchyma is located on the adaxial side of the stem and the chlorenchyma on the abaxial side, suggesting an homology with the anatomy of bifacial or dorsiventral mesophytic leaves for which the aerenchyma resembles spongy parenchyma (Mauseth 1991: 146–147). The abundance of intercellular air spaces may be an important adaptation to facilitate gas exchange and photosynthesis (Sajeva and Mauseth 1991). However, for *S. queretaroensis* the tissue may have evolved in response to the conditions of shade that occur during early stages of development in natural populations, as occurs for other species (Sifton 1957).

Physiology

Net CO_2 uptake by *S. queretaroensis* occurs primarily at night (Nobel and Pimienta-Barrios 1995), as is characteristic of Crassulacean acid metabolism (CAM) plants (Nobel 1988: 51–57). This behavior favors the ecological success of cacti in arid environments because less water is lost than for the species having stomatal opening during the daytime (Nobel 1995). Indeed, CAM is an extremely important adaptation that allows plants to successfully compete in environments with limited water or CO_2 availability (Cushman and Bohnert 1997).

As for laboratory studies (Nobel and Pimienta-Barrios 1995), gas exchange measurements in the field revealed that nearly all of the net CO_2 uptake for *S. queretaroensis* occurs at night. Young plants (three to four years old) had a higher maximum rate of CO_2 uptake (11.2 μmol m^{-2} s^{-1}) than adult plants (7.6 μmol m^{-2} s^{-1}). The total daily net CO_2 uptake, obtained by integrating the net CO_2 uptake rates over 24-h pe-

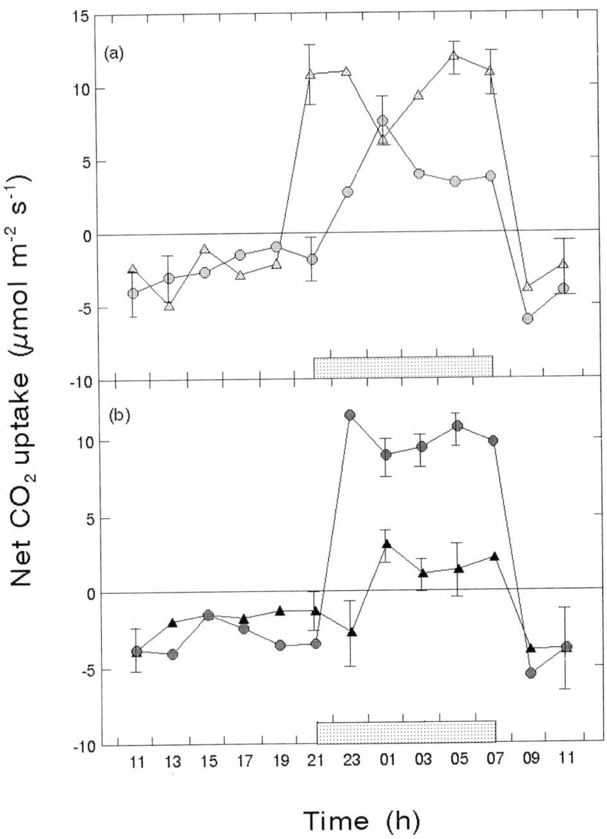

Fig. 5. Net CO_2 uptake by *S. queretaroensis* for young plants (△) and old plants (○; a), and light exposed stems (●) and shaded stems (▲; b) in the Sayula Basin, Jalisco, Mexico. Data are means ± SE (n = 12 stems). Previously unpublished data were obtained on 29–30 June 1997.

riods, was 30 mmol m^{-2} day^{-1} for the old plants and 303 mmol m^{-2} day^{-1} for the young plants (Fig. 5a). The young plants in the field had similar net CO_2 uptake as for *S. queretaroensis* of about the same age in a glasshouse (Nobel and Pimienta-Barrios 1995).

The rate of net CO_2 uptake depends on light level and hence varies with the orientation of the stems and their rib surfaces. Net CO_2 uptake increases more-or-less linearly with photosynthetic photon flux (PPF, wavelengths of 400 to 700 nm). Daily net CO_2 uptake over a 24-h period is 80 mmol m^{-2} d^{-1} for a north-facing surface (with a total daily PPF of 3.5 mol m^{-2} day^{-1}), 148 mmol m^{-2} day^{-1} for one facing northeast (PPF of 9.2 mol m^{-2} day^{-1}), 213 mmol m^{-2} day^{-1} for one facing south (PPF of 13.9 mol m^{-2} day^{-1}), and 317 mmol m^{-2} day^{-1} for one facing west (PPF of 19.2 mol m^{-2} day^{-1}; Nobel and Pimienta-Barrios 1995).

Similar tendencies for responses to PPF were observed in the field. Exposed stems receiving a total daily PPF of 16 mol m^{-2} day^{-1} during a partially cloudy summer day had 8 h of nocturnal CO_2 assimilation and a high maximal rate of net CO_2 uptake (11.5 μmol m^{-2} day^{-1}), whereas shaded stems receiving a PPF of 3 mol m^{-2} day^{-1} had 5 h of nocturnal CO_2 assimilation and a lower maximal rate (3.1 μmol m^{-2} s^{-1}; Fig. 5b). Net CO_2 uptake over 24 h was 300 mmol m^{-2} day^{-1} for the exposed stems and −32 mol m^{-2} day^{-1} for the shaded stems (Fig. 5b).

Temperature and drought duration affect net CO_2 uptake for glasshouse-grown plants under moderate total daily PPF (Nobel and Pimienta-Barrios 1995). The total daily net CO_2 uptake is unchanged as the minimal nighttime temperatures is reduced from 18°C to 13°C but decreases 25% when the minimal temperature is reduced to 8°C. Drought lasting 14 days does not significantly affect total daily net CO_2 uptake, but uptake is 33% lower after 27 days of drought, 69% lower after 41 days, and 89% lower after 56 days (Nobel and Pimienta-Barrios 1995).

Daily net CO_2 uptake for *S. queretaroensis* can be 340 mmol m^{-2} day^{-1} under wet conditions during the summer. Under similar conditions, daily net CO_2 uptake ranges from 158 to 285 mmol m^{-2} day^{-1} for four perennial CAM species native to the Chihuahuan or the Sonoran deserts and from 760 to 1170 mmol m^{-2} day^{-1} for five, highly-productive, cultivated, CAM species (Nobel 1988: 94–99). Thus even though *S. queretaroensis* is cultivated, its net CO_2 uptake more closely resembles that of wild perennial CAM species, probably reflecting its rather recent domestication and the fact that selections made by growers were based on fruit quality, phenologies, and suitability to adverse conditions rather than on biomass productivity (Nobel and Pimienta-Barrios 1995).

The absence of reproductive growth and the low vegetative growth during the wet summer for *S. queretaroensis,* coupled with the favorable environmental conditions for photosynthesis that prevail then, allows its photosynthetic stems to replenish and store carbohydrates and its root system to extend in order to absorb more water and mineral nutrients from the soil. This seasonal storage of carbohydrates, minerals, and water in its stems during the summer is essential for the ecological success of *S. queretaroensis.* Vegetative growth during the autumn and reproductive growth in the spring are supported by such reserves, enabling it to grow when soil water is scarce.

Similar to some conifers (Salisbury and Ross 1992: 258–259), juvenile plants of *S. queretaroensis* are sensitive to high light levels. When seedlings and cuttings are established in open fields, they usually become chlorotic and die. In contrast, both seedlings and cuttings of *Opuntia* spp. perform satisfactorily in the open, and they do not show the symptoms of high light damage that are evident for *S. queretaroensis.* This sensitivity to light is also revealed at the level of pigment stability. For instance, chlorophyll in *S. queretaroensis* is more sensitive to photooxidation (De Luna 1996) than that of *Opuntia* spp. (Neri-Luna, Pimienta-Barrios, and Robles-Murguia 1993).

Maximal rates and daily net CO_2 uptake by *S. queretaroensis* are higher under a doubled atmospheric CO_2 concentration of 720 μmol mol^{-1}, expected by the end of the twenty-first century, than under the current ambient atmospheric CO_2 concentration (Nobel 1996). Net CO_2 exchange becomes positive in midafternoon under the doubled CO_2 concentration. Because the total daily net CO_2 uptake is 36% higher for the doubled than the current atmospheric CO_2 concentration, *S. queretaroensis* will likely respond favorably as the atmospheric CO_2 concentration increases in the future, as for other CAM plants, expanding the regions where they may be profitably cultivated (Nobel 1996).

Mineral Nutrition

Under conditions of cultivation *S. queretaroensis* thrives in alluvial soils is in a relatively rich nutrient environment. Indeed, growers claim that cultivated plants show a negative response to the application of fertilizer, possibly due to low absorption rates (Pimienta-Barrios and Nobel 1994). Such responses to minerals may reflect its ecological origin, inasmuch as *S. queretaroensis* evolved in a low-nutrient environment. Plants growing in infertile habitats often enhance nutrient intake through a high root: shoot ratio and mycorrhizal associations (Chapin 1980). Although *S. queretaroensis* does not have a high root:shoot ratio, it has mycorrhizal associations (Arceta 1997). Mycorrhizae allow plants to expend less energy in foraging for soil resources, because it is energetically less costly to support the mycorrhizal fungi than to develop a high root:shoot ratio.

For *S. queretaroensis* the mutualistic association between the fungi and the roots is temporal, restricted to the rainy summer when fine roots (rain roots) develop on main roots. The fine roots become suberized and develop into permanent roots at the beginning of the autumn, when the mutualistic cooperation disappears. Thus the absorption of water and mineral nutrients occurs more readily during the summer wet season (Arceta 1997). The fine roots are superficial, and their association with mycorrhizae facilitates the absorption of water and minerals, even with low amounts of rainfall. For instance, light rains of only 6–7 mm on dry soils are enough for the absorption of water by roots of cacti (Jordan and Nobel 1982). However, the development of fine roots by *S. queretaroensis* is not strictly rain-induced, because they did not develop during the abnormally wet spring of 1997.

The levels of most elements in the chlorenchyma and the water-storage parenchyma of *S. queretaroensis* are similar to levels observed for other species of cacti (Nobel 1988: 176–180; Nobel and Pimienta-Barrios 1995). However, the N content in its chlorenchyma is only 7.8 mg g^{-1} dry weight, which is less than for any of the other 11 cactus species considered (average of 16.7 mg g^{-1}). Also, the macronutrient Ca averages only 6 mg g^{-1} in *S. queretaroensis* compared with 44 mg g^{-1} in the chlorenchyma of the other species, many of which contain prominent calcium oxalate crystals. The contents of the micronutrients Fe and Mn in the chlorenchyma of *S. queretaroensis* average 54% lower than the average for the other 11 species of cacti (Nobel 1988: 176–180). The relatively low contents of these two micronutrients and the extremely low content of nitrogen are consistent with the relatively low growth rates and low CO_2 uptakes rates for *S. queretaroensis* compared with highly productive cultivated CAM species (Nobel 1988; Nobel and Pimienta-Barrios 1995).

Seasonal Variation in Carbohydrate Content

Although seasonal variations of carbon accumulation and distribution have received little attention for CAM plants, it is widely accepted that starch is the major storage polysaccharide (Sutton, Ting, and Sutton 1981) that supplies carbon for malic acid synthesis during dark CO_2 fixation (Nobel 1994: 108–114). However, the role of mucilage in CAM plants is unclear. Some have discounted the role of mucilage in the water economy of cacti (Ting 1994), while others have demonstrated that the relative capacitance of the water-storage parenchyma is related to the mucilage content (Nobel, Cavalier, and Andrade 1992). Mucilage may serve both as a form of carbon storage and as a high molecular weight polymer that facilitates adaptation to drought and to low temperatures (Zañudo 1998).

Compounds with multiple physiological roles are common in plants, and the extent to which a compound serves a storage role depends on the environment (Chapin, Schulze, and Mooney 1990). For *S. queretaroensis* both starch and mucilage tend to increase in the stems at the beginning of the summer wet season, decreasing at the end of the summer and early fall (Fig. 6a,b). On the other hand, the content of reducing sugars in its stems is low at the beginning of the summer, increasing from August to September when both mucilage and starch decrease (Fig.6c). The storage of carbohydrates in *S. queretaroensis* during early summer coincides with favorable environmental conditions for photosynthesis, when carbon gains exceed demands for growth. In particular, root formation is the prominent morphological change then (Fig. 1a).

The seasonal variations in reserve carbohydrates and reducing sugars in the stems of *S. queretaroensis* are closely related to the beginning of the main vegetative and reproductive phenophases (Fig. 1a). Its seasonal cycles in carbohydrate accumulation are like those of deciduous plants (Chapin, Schulze, and Mooney 1990; Kozlowski, Kramer, and Pallardy 1991: 57) but contrast with woody perennials, in which carbohydrate content in branches reaches a maximum in autumn, decreasing rapidly in early spring (Haddad, Clair-Maczulajtys, and Bory 1995). The latter difference evidently results from both vegetative and reproductive phenophases in woody perennials occurring during spring and summer (Kozlowski, Kramer, and Pallardy 1991: 13–15), whereas for *S. queretaroensis* vegetative development and reproductive development do not coincide, as its stem extension occurs during late summer and autumn but reproductive growth occurs during late winter and spring (Pimienta-Barrios and Nobel 1995).

Conclusions

Stenocereus queretaroensis has various strategies of adaptation to aridity and at various levels—development, structure, and physiology. Reproductive development is asynchronous and does not overlap with vegetative development. Seeds mature at the end of the spring, favoring seed germination at the beginning of the summer rainy season, which leads to seedling establishment under advantageous soil moisture conditions when natural vegetation can act as nurse plants. Stem succulence allows the accumulation of water and carbohydrate reserves. Also, the photosynthetic stems have a low stomatal density and thick cuticle, restricting water loss. However, the most crucial adaptive trait to aridity is CAM, which greatly reduces water loss during CO_2 fixation.

The nocturnal fixation of CO_2 and the storage

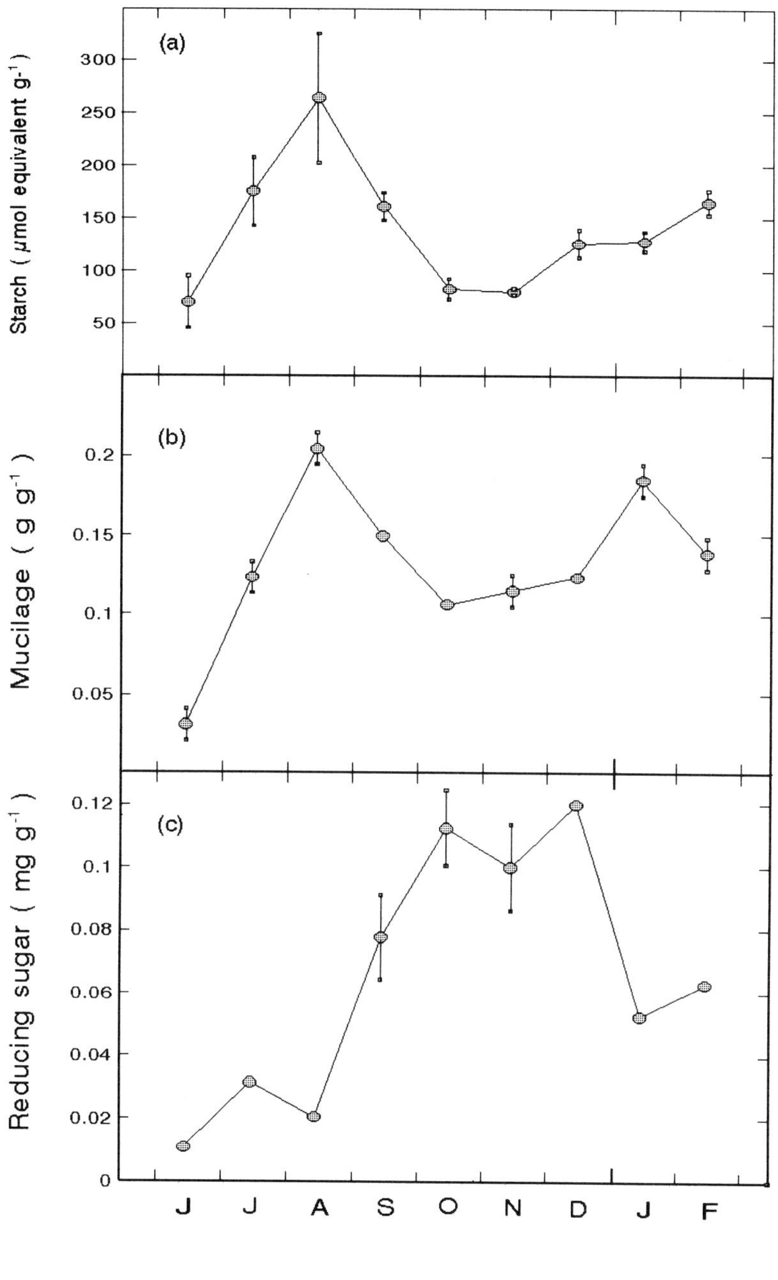

Fig. 6. Monthly levels of starch (a), mucilage (b), and reducing sugars (c) for *S. queretaroensis* in the Sayula Basin, Jalisco, Mexico. Data are on a dry weight basis and are means ± SE (n = 17 plants). Previously unpublished data were obtained during 1996–1997.

of carbohydrates, water, and minerals in the stems of *S. queretaroensis* during the summer, favored by the absence of reproductive growth and the relatively low stem extension then, is essential for its success in a subtropical semi-arid environment. The low rates of growth and photosynthesis for cultivated *S. queretaroensis* are associated with low tissue levels of nitrogen, chlorophyll, and some micronutrients (Fe, Mn), and perhaps low levels of gibberellic acid. Slow growth allows the plants to persist in arid and semi-arid environments, in which adverse environmental conditions must periodically be overcome.

ACKNOWLEDGMENTS

This project was financially supported by the Universidad de Guadalajara, CONACYT grant 0568P-B9506, and the UCLA Council on Research. We thank Celia Robles-Murguia, Lucila Mendez Moran, Alejandro Dominguez de la Torre and Enrico Yepez Gonzalez for field and laboratory assistance.

LITERATURE CITED

Arceta, E. G. I. 1997. Caracterización fisiológica de una población silvestre de pitayo (*Stenocereus queretaroensis* (Weber) Buxbaum) asociada con micorrizas vesículo arbuscular. Bachelor's Thesis. Universidad de Guadalajara, México.

Barbera, G. F., F. Carimi, and P. Inglese. 1992. Physical, morphological and chemical changes during fruit development and ripening in three cultivars of prickly pear, *Opuntia ficus-indica* (L.) Miller. Journal of Horticultural Science 67:307–312.

Bravo, H. H. 1978. Las cactáceas de México. Universidad Nacional Autónoma de Mexico, México, City.

Chapin, S. F. 1980. The mineral nutrition of wild plants. Annual Review of Ecology and Systematics 21:423–447.

———, **E.-D. Schulze, and H. A. Mooney.** 1990. The ecology and economics of storage in plants. Annual Review of Ecology and Systematics 21: 423–447.

Coombe, B. G. 1976. The development of fleshy fruits. Annual Review of Plant Physiology 27:507–528.

Cushman, C. J., and H. J. Bohnert. 1997. Molecular genetics of crassulacean acid metabolism. Plant Physiology 113:667–660.

De Luna, F. A. 1996. Efecto de la aplicación de inhibidores de la expresión genética sobre la senescencia de segmentos de tallo de pitayo (*Stenocereus queretaroensis*). Bachelor's Thesis. Universidad de Guadalajara, Mexico.

Esau, K. 1977. The anatomy of seed plants, 2nd ed. Wiley, New York.

Faust, M. 1989. Physiology of temperate zone fruit trees. Wiley, New York.

Gibson, A. C., and P. S. Nobel. 1986. The cactus primer. Harvard University Press, Cambridge, Massachusetts.

Grime, P. J. 1979. Plant strategies and vegetation processes. Wiley, New York.

———, **and R. Hunt.** 1975. Relative growth rate: its range and adaptive significance in local flora. Journal of Ecology 63:393–422.

Haddad, Y., D. Clair-Maczulajtys, and G. Bory. 1995. Effects of curtain-like pruning on distribution and seasonal patterns of carbohydrate reserves in plane (*Platanus acerifolia* Wild) trees. Tree Physiology 15:135–140.

Izquierdo, J., and J. Oltremari. 1996. Conservacion y uso sostenible de la biodiversidad en zonas áridas y semiáridas de América Latina y el Caribe. Pages 1–12 *in* Simonetti J. ed., Organización de las Naciones Unidas para la Agricultura y la Alimentación. Santiago, Chile.

Jimenez-Lopez, G. M., E. Pimienta-Barrios, and A. Muñoz-Urias. 1995. Estudio anatomico del tallo de pitayo (*Stenocereus queretaroensis* (Weber) Buxbaum). Revista Cactáceas y Suculentas Mexicanas 40:51–58.

Jordan, P. W., and P. S. Nobel. 1982. Height distribution of two species of cacti in relation to rainfall, seedling establishment, and growth. Botanical Gazette 143:511–517.

Kozlowski, T. T., P. J. Kramer, and S. G. Pallardy. 1991. The physiological ecology of woody plants. Academic Press, San Diego.

Larcher, W. 1995. Physiological plant ecology, 3rd ed. Springer-Verlag, Berlin.

Lee, S. K., and R. E. Young. 1983. Growth measurements as an indication of avocado maturity. Journal of the American Association for Horticultural Science 108:395–397.

Lomeli, M. E., and E. Pimienta. 1993. Demografía reproductiva del pitayo (*Stenocereus queretaroensis* (Weber) Buxbaum). Revista Cactáceas y Suculentas Mexicanas 38:13–19.

Mauseth, J. D. 1991. Botany: an introduction to plant biology. Saunders, New York.

Mizrahi, Y., A. Nerd, and P. S. Nobel. 1997. Cacti as crops. Horticultural Reviews 18:291–319.

Nerd, A., E. Raveh, and Y. Mizrahi. 1993. Adaptation of five columnar species to various conditions in the Negev Desert of Israel. Economic Botany 47: 304–311.

———, **and Y. Mizrahi.** 1997. Reproductive biology of cactus fruit crops. Horticultural Reviews 18: 321–346.

Neri-Luna, C., E. Pimienta-Barrios, and C. Robles-Murguia. 1993. Cambios fisiológicos durante el proceso de senescencia en cladodios jóvenes de nopal (*Opuntia ficus-indica* (L.) Miller). Agrociencia 4:7–18.

Nobel, P. S. 1983. Nutrient levels in cacti—relation to nocturnal acid accumulation and growth. American Journal of Botany 70:1244–1253.

———. 1988. Environmental biology of agaves and cacti. Cambridge University Press, New York.

———. 1994. Remarkable agaves and cacti. Oxford University Press, New York.

———. 1995. Environmental biology. Pages 36–48 *in* Barbera, G., P. Inglese, and E. Pimienta-Barrios eds., Agro-ecology, cultivation and uses of cactus pear. FAO, Rome.

———. 1996. Response of some North American CAM plants to freezing temperatures and doubled CO_2 concentrations: implications of global climate change for extending cultivation. Journal of Arid Environments 34:187–196.

———, **J. Cavalier, and J. L. Andrade.** 1992. Mucilage in cacti: its apoplastic capacitance, associated solutes, and influences of tissue water relations. Journal of Experimental Botany 43:641–648.

———, **and E. Pimienta-Barrios.** 1995. Monthly stem elongation for *Stenocereus queretaroenesis*: relationships to environmental conditions, net CO_2 uptake, and seasonal variations in sugar content. Environmental and Experimental Botany 35:17–42.

Osuna, E. T., and E. Pimienta. 1992. Diferenciación floral en aguacate. Agronomía 1:19–21.

Perez, R. C. 1993. Viabilidad de semillas y poliembrionía en morfoespecies cultivadas y silvestres de nopal tunero (*Opuntia* spp.). Bachelor's Thesis. Universidad de Guadalajara, Guadalajara, Mexico.

Pimienta, E. B. 1990. El nopal tunero. Universidad de Guadalajara, Guadalajara, Mexico.

———, **M. M. Loera, and L. O. Lopez**. 1993. Estudio anatomico comparativo entre colectas del subgenero *Opuntia.* Agrociencia 4: 7–21.

Pimienta-Barrios, E., and P. S. Nobel. 1994. Pitaya (*Stenocereus* spp., Cactaceae): an ancient and modern fruit crop of Mexico. Economic Botany 48:76–83.

———, **and P. S. Nobel.** 1995. Reproductive characteristics of pitayo (*Stenocereus queretaroensis*) and their relationships with soluble sugars and irrigation. Journal of the American Society for Horticultural Science 120:1082–1086.

———, **E., C. Robles-Murguia, and A. Dominguez de la Torre.** 1995. Estrategias fisiologicas y reproductivas de adaptación del pitayo a la aridez. Ciencia 46:339–349.

———, **G. Hernandez, A. Dominguez, and P. S. Nobel.** 1998. Growth and development of the columnar cactus *Stenocereus queretaroensis* (Weber) Buxbaum in a subtropical environment, including effects of gibberellic acid. Tree Physiology 18:59–64.

Robles, M. C. 1994. Estudio anatómico-fisiológico comparativo entre el nopal (*Opuntia ficus-indica* (L.) Miller) y el pitayo (*Stenocereus queretaroensis* (Weber) Buxbaum). Bachelor's Thesis. Universidad de Guadalajara, Mexico.

Ryugo, K. 1988. Fruit culture. Wiley, New York.

Salisbury, B. F., and C. W. Ross. 1992. Plant physiology, 4th ed. Wadsworth, Belmont, California.

Sajeva, M., and J. D. Mauseth. 1991. Leaf-like structure in the photosynthetic, succulent stems of cacti. Annals of Botany 68:405–411.

Sifton, H. B. 1957. Air-spaces in plants. Botanical Review 11:108–143.

Sutton, B. G., I. P. Ting, and R. Sutton. 1981. Carbohydrate metabolism of cactus in a desert environment. Plant Physiology 68:784–787.

Ting, I. P. 1994. Carbohydrate metabolism in cacti: gums and mucilage. Pages 7–12. *in* P. Felker, ed., Proceedings of the 5th Annual Texas Prickly Pear Council. Kingsville, Texas.

Zanchin, A. C., C. Bonghi, G. Casadoro, A. Ramina, and N. Rascio. 1994. Cell enlargement and cell separation during peach fruit development. International Journal of Plant Sciences 155:49–56.

Zañudo, H. J. 1998. Relación de los factores climáticos con el desarrollo y la actividad fisiológica en poblaciones silvestres y cultivadas de pitayo (*Stenocereus queretaroensis* (Weber) Buxbaum). Bachelor's Thesis. Universidad de Guadalajara, Mexico.

XIXICAMÁTIC OR WOOD ROSE (*MERREMIA TUBEROSA*, CONVOLVULACEAE): ORIGINS AND DISPERSAL[1]

DANIEL F. AUSTIN

Austin, Daniel F. *(Department of Biological Sciences, Florida Atlantic University, 777 Glades Road, Boca Raton, FL 33431).* XIXICAMÁTIC OR WOOD ROSE (*MERREMIA TUBEROSA*, CONVOLVULACEAE): ORIGINS AND DISPERSAL. *Economic Botany 52(4): 412–422. 1998. The nativity of* Merremia tuberosa *is examined with historical, ecological, linguistic, cladistic, and biogeographic data. All data indicate that the species was originally from southern Mexico and nearby Central America. The closest relatives of* M. tuberosa *are* M. austinii *and* M. discoidesperma. Merremia tuberosa *was first recorded from Mexico as* xixicamátic *in the middle 1500s by Fr. Bernadino Sahagún; no connection was made between that plant and the Convolvulaceae by subsequent authors. Plukenet later recorded* M. tuberosa *in Jamaica in 1691. By 1731 or earlier* M. tuberosa *was cultivated at the Chelsea Physic Garden, London, England. Soon afterward, wood rose had been spread around the world and grown for medicine. The roots contain resins that formerly were used across the tropics and in Europe as laxatives. Now plants are grown for their flowers and ornamental fruits that are used by florists.*

XIXICAMÁTIC Ó WOOD ROSE (*MERREMIA TUBEROSA*, CONVOLVULACEAE): ORIGENES Y DISPERCIÓN. *El origen de* Merremia tuberosa *es examinado por medio de consideración de datos históricos, ecológicos, lingüísticos, cladísticos y biogeográficos. Los resultados indican que la especie es originalmente del sudeste de Méjico y cercanias de Centro-america. Los relativos mas cercanos de* M. tuberosa *son* M. austinii *y* M. discoidesperma. *Merremia tuberosa fue inicialmente documentada como xixicamátic a mediados de 1500 por Fr. Bernardo Sahagún. Ninguna conexión fue hecha entre la planta y convolvulaceas por autores subsiguientes. Mas tarde Plukenet documentó* M. tuberosa *en Jamaica en 1691. En 1731 ó antes,* M. tuberosa *fue cultivada en el "Chelsea Physic Garden" Londres, Inglaterra. Pronto después, wood rose fue esparcida alrededor del mundo y cultivada con propósitos medicinales. Las raices contienen resinas que anteriormente fueron usadas como laxantes atravez de los trópicos y Europa. Hoy en día, las plantas son cultivadas por sus flores y frutas ornamentales usadas por los floristas.*

Key Words: *Merremia tuberosa,* medicinal plant, laxative, Brazilian jalap, cladistic analysis, biogeography, human transport, wood rose.

Little comparative examination of the early Spanish reports of medicinal plants has been made except for species that became economically important and famous, like chile (*Capsicum annuum* L.), guayacán (*Guiacum officinale* L.), and tobacco (*Nicotiana tabacum* L.). Yet, the early Spanish visitors in the New World, especially physicians and priests, were astonished with the knowledge people had of plants. Some, like Fr. Bernardino de Sahagún (Garibay 1985), G. F. de Oviedo (1526), Nicolás Monardes (1569), and Francisco Ximénez (1615) recorded what native Americans know about medicinal uses of plants.

Schultes (1941) was one of the first to identify as *Turbina corymbosa* (L.) Raf. (=*Rivea corymbosa* (L.) Hall. f.) the seeds Náhuatl speakers of Mexico called *ololiuhqui.* However, the Spanish documents make it clear that Mexicans recognized several kinds of *ololiuhqui.* Schultes and Hofmann (1979) identified all but one of those species. The *ololiuhqui* that they did not identify was also called *xixicamátic* by Sahagún in the middle 1500s (Garibay 1985). The following discussion suggests that *xixicamátic* is *Merremia tuberosa,* and examines the discovery and spread of this medicinal plant around the world.

[1] Received 21 October 1997; accepted 7 June 1998.

Historical Discovery and Dispersal

Merremia tuberosa first became known to the European scientific community when recorded by Plukenet (1691, 1696) and Sloane (1696, 1707). Plukenet apparently based his report on specimens sent to him. Sloane based *Convolvulus major heptaphyllus* on his own Jamaica collections made between 1687 and 1689.

At least as early as 1731 *M. tuberosa* was in cultivation in the Chelsea Physic Garden, London (Miller 1768). Seventeen years later an international seed exchange resulted in the *Merremia* being cultivated in Europe as far north as the Uppsala Botanic Garden, Sweden (Linnaeus 1748). Soon afterward Browne (1756) gain recorded it in Jamaica. *Merremia tuberosa* continued being cultivated in London in the late 1760s (Miller 1768).

More seeds were taken to Europe in 1793 from Jamaica (Powell 1977). This later introduction was brought on the return voyage of the *H.M.S. Providence. Merremia tuberosa* was among the species carried because they were "...rare plants at home [England]" (Powell 1977, quoting from Dr. Broughton's 1793 list). *Merremia* was still in cultivation in the early 1800s when it was illustrated for the third time (Anonymous 1807). These vines were grown at Kew, near London (Aiton 1810), and finally flowered there in 1822 (Edwards 1823). *Merremia tuberosa* was known from plants cultivated in northern Europe for at least 88 years before it was first recorded in Cuba (Humboldt, Bonpland and Kunth 1819).

By the early 1800s *M. tuberosa* also had been taken to Africa (Welwitsch 1858), Mauritius (Baker 1877; Bojer 1837), Hong Kong (Hemsley 1892), the Philippines (Blanco 1837), Ethiopia, Java and Australia (Miquel 1857). Meisner (1869) had seen specimens from Mexico, the Caribbean, the Guianas, Asia, and Africa, but not Brazil. Hallier (1897) recorded the first specimens of *M. tuberosa* from Brazil.

Neither Gray (1878, 1886) nor Chapman (1897) recorded *M. tuberosa* in Florida; Small (1913) and Simpson (1916) first found it there. Thus, it was introduced relatively late into the USA. *Merremia tuberosa* apparently also arrived late in Hawaii, with the first collection having been made in 1932 (Austin 1990). *Merremia tuberosa* is just now arriving on some of the islands in the Pacific (e.g., Rarotonga, Cook Islands, William R. Sykes, pers. comm. Feb. 1996).

Taxonomy

During the years between 1691 and 1753, the vines were called a *Convolvulus*; various phrase-names were used. Subsequently, Linnaeus (1753) took up one of the descriptors applied by Plukenet and the plants became *Ipomoea tuberosa.* Finally, the name became *Merremia tuberosa* (L.) Rendle (1905).

Through most of the 1800s people thought that *M. tuberosa* was native to tropical Africa, Asia and the Americas (Choisy 1845; Grisebach 1864; Hemsley 1882; Meisner 1869). This confusion occurred because these authors combined an American with an Old World species under a single name. Clarke (1883) finally called the African plants *Ipomoea kentrocaulos*; later this became *Merremia kentrocaulos* (Steudel ex C. B. Clarke) Rendle.

To make the situation more complicated, *M. tuberosa* became confused with a drift seed known from at least the late 1500s in northern Europe. The drift seeds, first illustrated by Clusius in 1605, are now known as *Merremia discoidesperma* (Donn.-Sm.) O'Donell. Williams (1970a) republished the Clusius illustration along with its accompanying text. Soon afterward, Gunn (1977) gave a complete discussion of this confusion, including comments on the paper about "*Ipomoea tuberosa*" by Hemsley (1892) that had illustrations of both species under the single name. *Merremia tuberosa* is a rare drift seed.

Merremia tuberosa (L.) Rendle in Thiselton-Dyer, Fl. Trop. Afr. 4:104. 1905.

Ipomoea tuberosa L., Sp. Pl. 160. 1753; O'Donell, Lilloa 6:514. 1941. Type: Jamaica. herb. Linnaeus (LINN 219.4, microfiche seen). Selected lectotype by Austin (1975). Comments from N. Turland (in litt. 8 Jan. 1997) on LINN specimen: "The writing in the bottom right is 'tuberosa JES' by James Edward Smith, not 'I. tuberosa' as Savage (1945) said. At the base of the stem is '5' (the species number of *I. tuberosa* in *Species plantarum*), 'digitata' and 'tuberosa', all by Linnaeus, with the 'tuberosa' also deleted by Linnaeus (not the 'digitata' deleted, as Savage said). There is no known provenance or collector. We can only deduce that it was in Linnaeus's possession prior to 1753 because he wrote the *Species plantarum* number on the sheet."

Convolvulus major heptaphyllus, flore sulphureo, odorato, speciocissimo. Sloane, Jamaica 55. 1696; Hist. 1:152. t. 96. f.2. 1707. Sloane further indicates "Spanish arbor vine, or Spanish woodbine. Inter arbores prope ripas rio Cobre non procul ab urbo St. Jago de la Vega provenit."

Convolvulus americanus, mandjuccae multifido folio, heptaphyllos, flore albo fundo purpureo, radice tuberosa, cortice albo Plukenet, Almagestum 116. 1696; Phytographia t. 276. f. 6. 1691. Based on Herb. Sloane vol. 95, fol. 188 verso (BM). Although I have not seen this specimen, N. Turland at the British Museum (in litt. 1996) tells me that it is clearly the same as the drawing published in 1691. The flower color description is in error.

Convolvulus gossypiifolius Humboldt, Bonpland and Kunth, Nov. Gen. Sp. Pl. 3:107. 1819. Type: Cuba. prope Havana. *Bonpland* (microfiche seen).

Convolvulus tuberosus (L.) Sprengel, Syst. Veg. 1:591. 1825.

Convolvulus macrocarpus Sprengel, Syst. Veg. 1:592. 1825, non L. (1759).

Batatas tuberosus (L.) Bojer, Hort. Maurit. 226. 1837.

Ipomoea mendesii Welw., Apont. 584. 1858. Type: Angola. Benguella, in thickets and introduced into gardens in Loanda. *Welwitsch 6254* (BM fide Verdcourt 1963).

Operculina tuberosa (L.) Meisner in Martius, Fl. Bras. 7:212. 1869.

Ipomoea nuda Peter in Engler and Prantl, Nat. Pflanzenfam. 4(3a):31. 1891, nomen nudum.

Ipomoea glaziovii Dammer, Bot. Jahrb. Syst. 23(Beibl. 57):40. 1897. Type: Brazil. Rio de Janeiro. prope Piratininga. *A. Glaziou 15267* (holotype B, destroyed?; isotype C fide Verdcourt 1963).

Taproots tuberous. **Stems** basally woody, perennial, twining, glabrous. **Leaves** simple, the blades ± circular in outline, 6–16 cm long and wide, the base cordate, margins palmately 5–7-lobed almost to the base, the lobes 8–20 cm long, 9–20 cm wide, ovate, 3–9 cm long, 1–5 cm wide, margins entire, the apex attenuate, the secondary venation pinnate, glabrous; **petioles** (3–)6–18 cm long, glabrous. **Cymes** axillary, with the first branch dichasial, the others monochasial, racemose, the bracts triangular, 2 mm long, caducous; **pedicels** 15–18 mm long, claviform, enlarging in fruit, glabrous; **sepals** unequal, the outer two longer than inner three, oval to almost orbicular, apex rounded, 23–25 mm long, membranous apically, somewhat herbaceous basally, the margins scarious, with sepals ± equally enlarged in fruit, the inner three oblong, 12–20 mm long; **corolla** contortiplicate, enclosed by the sepals in bud, funnelform, (4–) 5–6 cm long, yellow, glabrous, the 5 veins obscure; **filament** unequal, of three lengths, 2.5–3 cm long, the base dilated, glandular pubescent; **anther** dehiscence introrse, twisted, pollen spheroidal, diameter 75–86 μ, 12–15-rugate; **pistil** glabrous, 4-locular, the stigma globose. **Fruits** globose to depressed globose, 3–3.5 cm in diameter, the calyx accrescent, with fruiting sepals divergent but supporting the fruit, the pericarp chartaceous, tardily irregularly dehiscent, sometimes becoming somewhat circumscissile; **seeds** 1–4, ovoid, 1.5–2 cm long, the surface smooth, covered with short, erect, puberulent indumentum, glabrescent, hilum circular with a raised u-shaped margin, cotyledon germination hypogaeous. Distribution (Fig. 1).

Common Names—Confusion since the 1500s

Sahagún wrote sometime between 1547 and 1582: "There is another herb called *ololiuhqui*, or *xixicamátic,* it has leaves like *miltómate* [*Physalis* sp.], remotely placed, the flowers are yellow; the flowers are not used, nor the leaves or branches. It has a rounded and large rootlike the turnip (translated from Garibay 1985: book XI, ¶121, p. 678). *Xixicamátic* comes from *xixi* = belly and *camátl* = mouth, hence "from the mouth and belly."

This *ololiuhqui* called *xixicamátic* by Sahagún is a twiner (*ololiuhqui* has two meanings, "semilla redonda," or round seed, and "que dar vueltas," or twiner). If the assumption is made that it, like all other *ololiuhqui,* is one of the Convolvulaceae, it can only be one Meso-American species. The yellow flowers, tuberous root, and lobed leaf margins are distinctive. The description best matches *M. tuberosa.*

Merremia tuberosa has at least 42 common names, with southern Mexico and Guatemala contributing the most. Guatemala has the greatest variety of names recorded, including "bejuco

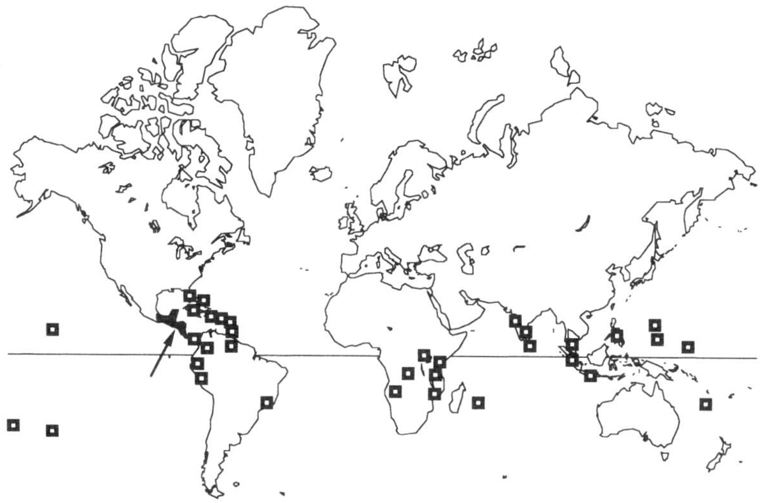

Fig. 1. Distribution of *Merremia tuberosa*. Points are documented by herbarium specimens. Plants are likely cultivated in other places (e.g., mainland China), but no specimens have been seen.

de golondrina," "foco de luz," "quiebra cajete," "quiebra machete," "quinamacal," and "rosa de barranco."

Now the most commonly used name in American English for *M. tuberosa* is "wood rose." This is the plant of the horticulture and florist trade. The name seems to have come into existence in the 1940s (Bailey 1949; Yuncker 1943); it was not used by Small (1913, 1933) or Bailey (1935).

Brazil's ancient use of *Operculina* roots as a laxative gave rise to several species in different genera being called "Brazilian jalap" in the 1800s. Indeed, *M. tuberosa* is not native in Brazil and is not mentioned by either LeCointe (1941) or Mors and Rizzini (1966). That Europeans still call *M. tuberosa* "Brazilian jalap" probably dates from that old confusion when several *Ipomoea, Operculina macrocarpa* (L.) S. Manso, and *O. pteripes* (G. Don) O'Donell, and other species were all called "jalap."

The name "jalap" has been used for purgative roots in Spanish since the early 1600s and in English since the 1670s (Corominas 1954; OED 1971; Ximénez 1615). "Jalap" in European English simply has denoted "laxative" from the 1600s to the present. There has been confusion since the 1500s between the various kinds of "jalap" originating in the Americas. McDonald (1982) identified four *Ipomoea* species (*I. jalapa* (L.) Pursh, *I. orizabensis* (Pelletan) Ledenois, *I. purga* (Wender.) Hayne, *I. simulans* Hanbury) that were called "jalapa." This confusion was compounded not only by indigenous uses of several species, but by the frequent adulteration of products at their source (Linajes, Rico-Gray and G. Carrión 1992; Williams 1970c). Some of the reports of laxative uses for "jalap" also include *M. tuberosa* (Shellard 1961). *Ipomoea, Merremia* and *Operculina* have all been added to the mixtures sold as "jalap."

Apparently *M. tuberosa* is uncommon in India. Hallier (1897) cited specimens from the Deccan, Bombay, and Madras; however, the species is not listed for India's Hassan District (Saldanha and Nicolson 1976), Tamilnadu Carnatic (Rani and Matthew 1983), Bangladesh (Kahn 1985), or West Pakistan (Austin and Ghazanfar 1979). Presumably the specimens cited by Hallier were from cultivation. *Merremia tuberosa* is cultivated in Sri Lanka, but has no common name there (Austin 1980). The species was introduced into India within the last 300 or so years; it has no name in Sanskrit.

No Chinese name is recorded; it is not listed for China or Taiwan (Chang 1978; Fang and Staples 1995). *Merremia tuberosa* is cultivated in southern China, but not established in the flora anywhere except Taiwan (G. W. Staples, pers. comm. Nov. 1996). The Japanese name "bara-asa-gao" (rose morning face) is used in the Ryukyu Islands (Walker 1976); that compound name is based on the one for *Ipomoea nil* (L.) Roth ("asa-gao"), an American species. Similarly, the species is not in southeastern Asia (Gagnepain and Courchet 1915) nor Thailand (Craib and Kerr 1954).

Uses and Chemistry

Sahagún wrote of *M. tuberosa* usage: ". . .the grated root is useful for those that have swollen bellies and whose intestines rumble; when drunk while fasting, it purges, and lowers fever . . . The root is somewhat sweet, and one makes enough to drink three times. . ." (translated from Garibay 1985: book XI, ¶121, p. 678). Soon after its description by Plukenet and Sloane, *M. tuberosa* was grown as a medicinal plant in London's Chelsea Physic Garden (Miller 1768).

Some of the current Spanish common names indicate a continuing medicinal usage as a laxative, i.e., "batatilla ventruda," "quiebra cajete." The twiner in Cuba was considered a "drástico" (Roig y Mesa 1945); they considered *Merremia* secondary to the other "aguinaldo rosado" (*Ipomoea indica* (Burm. f.) Merr.). Alcorn (1984) found the Huastecan Maya using *M. tuberosa* and calling it, among other names, "pak-tha' pok' laak" (big broken bowl). The "pok' laak" refers to *Operculina pinnatifida* (Kunth) O'Donell, which the Maya considered so powerful that just handling the plant will "easily break dishes at home" (Alcorn 1984). This concept doubtless is associated with the Spanish "quiebra cajete" for *M. tuberosa* and several other members of the family. The Maya use *M. tuberosa* against headaches (Alcorn 1984).

Those *Merremia* species that have been examined contain resins and a number of other compounds; these active ingredient are in both roots and seeds (Farnsworth et al. 1969). *Merremia tuberosa* has been reported to contain saponins (Hegnauer 1964); however, they have never been isolated. Prof. E. Eich (pers. comm. Dec. 1996) says that these chemicals are probably glycorhetins (resins) that give the same reaction in bioassays as saponins. In addition, Eich (pers. comm. 1996) and Schimming et al. (1996) provided additional information about the chemicals in *M. tuberosa.* Roots contain the lipophilic alkaloid tropan-3-one (=tropinone), calystegines, and several hygrine alkaloids (e.g., hygrine, cuscohygrine, 2′,4-N-methylpyrrolidinylhygrine and its 2′,3-isomer). In addition, there are the coumarins umbelliferon and scopoletin, common in the Convolvulaceae. These compounds are in epigeal parts and roots. These also are two hydroxycinnamic acid alkyl esters (hexadecyl *p*-coumarate and octadecyl *p*-coumarate). A flavonoid (quercetin) was also reported (Nair, Daniel and Sabuis 1984).

Where is the Species Native?

Ecology. The only place *M. tuberosa* is considered wild in native undisturbed habitats is southern Mexico and nearby Central America. In Mexico, Sahagún said ". . .it grows in the hills and level country" (translated from Garibay 1985: book XI, ¶121, p. 678). McDonald (1987, 1993) recorded the species as wild in Veracruz and nearby states in Mexico. There the plants grow in evergreen tall forest, median forest, low deciduous forest and river or thicket zones. Williams (1970b) recorded it in "moist or wet thickets" in Guatemala. In Costa Rica, Standley (1938) noted that the species was wild "in thickets and forests of the Atlantic tierra caliente" and cultivated for ornament.

Corresponding to this range is the distribution of the seed predator *Megacerus capreolus* (Jekel). Many beetles in *Megacerus* (Bruchidae) are obligate to single species of Convolvulaceae. Apparently, *M. capreolus* is one of those species because it has not been recorded on any member of the family except *Merremia tuberosa.* The range of this beetle may match the range of nativity of *M. tuberosa,* from Mexico to Panama (Teran and Kingsolver 1977).

Caribbean plants were originally recorded only from cultivation in Jamaica in the 1600s. Both Browne (1756) and Adams (1972) considered *M. tuberosa* a garden escape that had become partly naturalized and invasive. Since Hurricane Andrew hit Florida in 1992, the wild *M. tuberosa* populations have outgrown and overshadowed native vegetation (Gordon and Thomas 1994; Austin 1998).

The species is also considered an introduction for Puerto Rico (Liogier and Martorell 1982), Cuba (Leon and Alain 1974), and the Lesser Antilles (Fournet 1978; Powell and Staples 1989). Even when discovered by Humboldt, Bonpland and Kunth (1819), the plants were "growing in cultivation on the island of Cuba, near Havana" (crecit in cultis insulae Cuba, prope Havana).

Austin (1982a,b) recorded *M. tuberosa* in Venezuela and Ecuador only cultivated and escaped in thickets. *Merremia tuberosa* has only been collected in Peru from cultivation (Brako

TABLE 1. DATA MATRIX FOR *MERREMIA TUBEROSA* AND RELATED SPECIES. MISSING DATA ARE INDICATED BY A "?". MULTIPLE-CHARACTER STATES ARE: A = 0 & 1; B = 3 & 4. FOR CHARACTER NAMES AND STATES, SEE TABLE 3.

Characters	12345	1 67890	11111 12345	11112 67890	22222 12345	22223 67890	33 12
M. aurea	01110	10011	?0010	00001	???01	10???	?3
M. austinii	01011	1?A12	10?00	1011?	??010	01111	12
M. discoidesperma	00000	00010	10110	10110	??110	11111	12
M. dissecta	01111	10110	00001	01010	00001	00000	0B
M. kentrocaulos	01011	11111	10110	10???	10001	00???	?1
M. palmeri	01110	10010	?0011	10001	21001	110??	?3
M. platyphylla	01110	1??12	?0011	10011	??001	11???	?3
M. tuberosa	00010	10010	10110	10110	21110	11111	12
Op. turpethum	10000	01100	01000	10010	10020	00000	00
Op. hamiltonii	10000	01100	01000	11?21	1?020	00???	?2
Op. macrocarpa	11110	11000	?1000	110??	10120	?000?	04

and Zarucchi 1993). Austin and Cavalcante (1982) did not find the species in Amazonia.

In the Old World, the plants are uniformly recorded from cultivation, from thickets as a garden escape, or both (e.g., Backer 1965; Gonçalves 1987; Hallier 1894; Lejoly and Lisowski 1992; Meeuse 1957; Ooststroom 1953; Rendle 1905; Verdcourt 1963). Baker (1877) said that *M. tuberosa,* having been "spread round the world in the tropics, is naturalised in Mauritius."

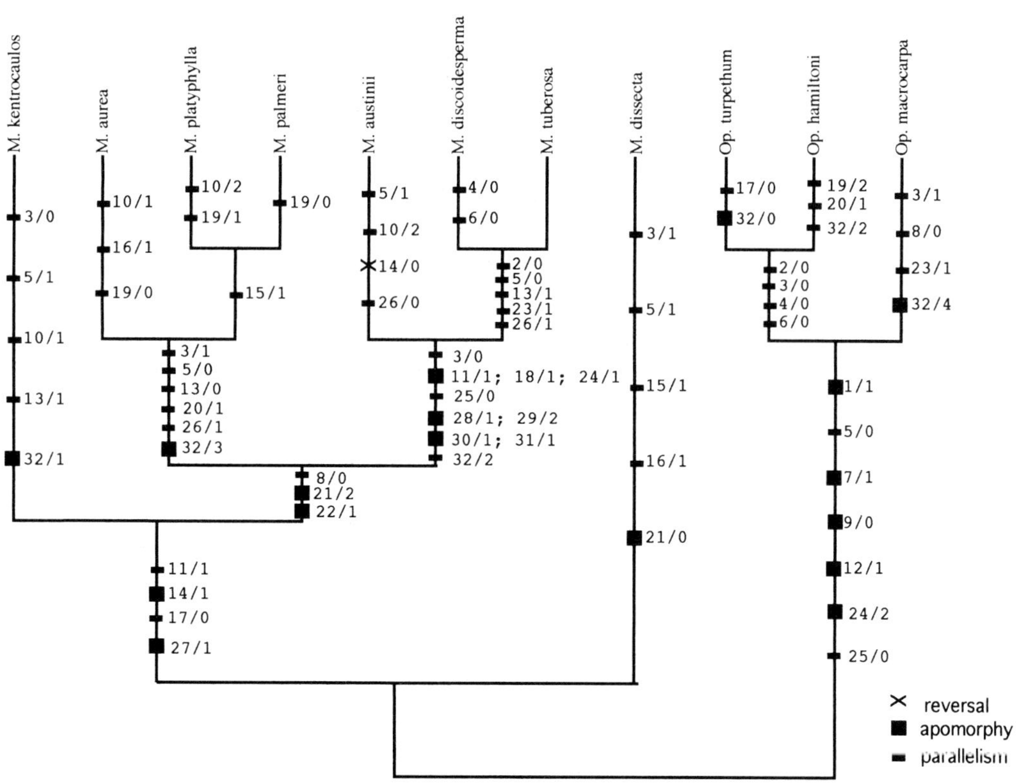

Fig. 2. Cladogram of probable relatives of *Merremia tuberosa.* Numbers along the internodes indicate the traits and states changing at those points (e.g., 25/0 means that trait 25 changed to state 0).

TABLE 2. CHARACTERS AND CHARACTER STATES USED FOR CLADISTIC ANALYSIS OF *MERREMIA TUBEROSA* AND RELATED SPECIES.

Characters	Character states
1. Stem wing:	0: absent; 1: present
2. Leaves circular to reniform	0: no; 1: yes
3. Leaf type	0: simple; 1: compound
4. Leaf lobe palmate	0: no; 1: yes
5. Leaf margin/leaf lobe	0: entire; 1: tooth or lobed
6. Leaf venation	0: pinnate; 1: palmate; 2: absent
7. Bract type	0: scale-like (≤5 mm); 1: leaf-like (≥10 mm)
8. Outer sepal tip	0: round; 1: acute to acuminate
9. Sepal shape	0: ovate to orbicular; 1: oblong-elliptic
10. Outer flowering sepals	0: longer; 1: shorter; 2: equal
11. Corolla aestivation	0: plicate; 1: contortiplicate
12. Interplical pubescence	0: glabrous; 1: pubescent
13. Interplical veins	0: distinct; 1: obscure
14. Corolla size	0: 2.5–4 cm; 1: >4 cm
15. Corolla colors	0: yellow; 1: white
16. Corolla pattern	0: light limb + dark throat; 1: uniform
17. Corolla shape	0: funnel; 1: bell
18. Filament length series	0: one; 1: three
19. Filament adnation	0: <0.33%; 1: 0.33–0.5%; 2: 0.66–0.75%
20. Anther length	0: 2–5 mm; 1: 7–12 mm
21. Pollen shape	0: oblate; 1: prolate; 2: spheroidal
22. Pollen size	0: medium (45–75μ); 1: large (≥80μ)
23. Capsule length	0: 1–2 cm; 1: >2 cm
24. Fruit dehiscence	0: capsular; 1: tardy; 2: opercular
25. Fruit valves	0: none; 1: four
26. Seed length	0: 4–9 mm; 1: 10–20 mm
27. Seed pubescence	0: glabrous; 1: puberulent
28. Cotyledon position	0: epigeal; 1: hypogeal
29. Cotyledon lobing	0: yes; 1: no
30. Cotyledon or Lobule shape	0: ovate; 1: absent
31. Basal sinus	0: present; 1: absent
32. Nativity	0: Asia; 1: Africa; 2: e Mexico & Central America; 3: w Mexico; 4: Caribbean

Relationships. Pollen and other data (Ferguson, Verdcourt and Poole 1977; Austin and Staples 1980) indicate that *Merremia* contains several distinct species-groups that might better be called separate genera or subgenera. Not all the species have been studied enough for formal taxonomic changes. Even though there are taxonomic problems within *Merremia,* the American species studied here form an apparently monophyletic group.

Merremia tuberosa is in sect. *Schizips* (Griseb.) O'Donell (1941). Mexican species in the same section (Tables 1,2) thought to be allied with *M. tuberosa* (McDonald 1987), and types of each of the other *Merremia* sections sensu Ooststroom (1953) were studied. The Old World *Merremia kentrocaulos* was included because it has been confused with *M. tuberosa.* Because analysis of all genera in the family (Austin 1997) shows *Operculina* to be the taxon most closely related to *Merremia,* three species (*O. hamiltonii* (G. Don) D. F. Austin and G. W. Staples, *O. macrocarpa* (L.) Urban, *O. turpethum* (L.) S. Manso) were used in the outgroup.

The data were analyzed with MacClade (Maddison and Maddison 1992) and PAUP (Swofford and Begle 1993). Preliminary analysis indicated that the species placed in *Merremia* sect. *Merremia* (*M. hederacea* (Burm. f.) Hall f.), *M.* sect. *Halliale* Hall. f. (*M. peltata* (L.) Merrill), *M.*

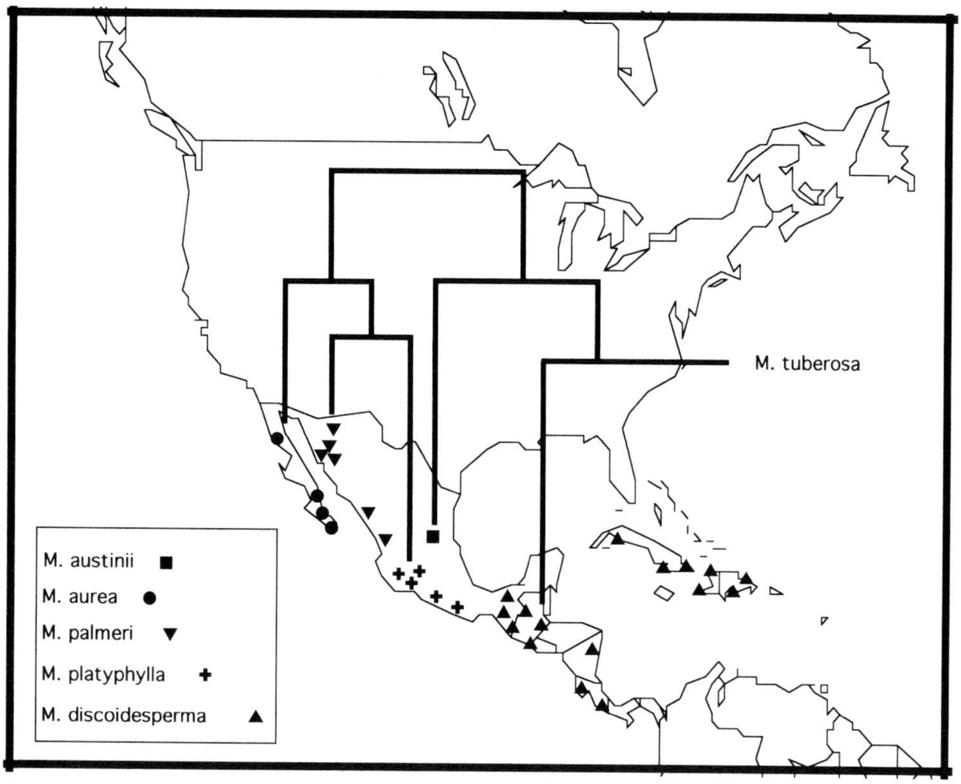

Fig. 3. Biogeography overlain on the cladogram. This analysis strengthens the conclusion of an American origin of *Merremia tuberosa* in southern Mexico and nearby Central America. *Merremia tuberosa* is left unmapped for comparison with the mapped species.

sect. *Wavula* Ooststroom (*M. similis* Elmer) and *M.* sect. *Xanthips* (Griseb.) Hall. f. (*M. umbellata* (L.) Hall. f.) were distinct from those more closely allied to *M. tuberosa.* These sections were excluded from further analyses.

The restricted group was analyzed in two ways: including traits 1, 7, 9, 12, and 32; and excluding these. Branch-and-bound searches used unordered traits, variable characters treated as polymorphisms, TBR branch swapping, and collapsing zero-length branches. The result using all 32 traits was two equally parsimonious trees having 62 steps; the consistency index is 0.66, and the homoplasy index 0.34. Exactly the same tree topology was found when trait 32 (nativity) was excluded. When traits 1, 7, 9, 12, and 32 were excluded, two trees resulted with a length of 52 steps, a consistency index of 0.62, and a homoplasy index of 0.38 (Fig. 2). The only differences between these two trees and those obtained with all 32 traits were that the clade containing *M. aurea, M. palmeri* and *M. platyphylla* was reduced to a polytomy.

Four of the relatives of *M. tuberosa* are endemic to Mexico. *Merremia discoidesperma* is a near-endemic to Central America, with disjunct populations in the Greater Antilles. Addition of biogeographic data to the matrix allows a comparison of the relationships with distribution (Fig. 3). These data indicate that *M. tuberosa* and all its close relatives are American.

All the data indicate that *M. tuberosa* is native to Mexico and nearby Central America near its relatives *M. austinii* and *M. discoidesperma. Merremia tuberosa* had been considered medicinal on mainland America since before the arrival of Europeans. Currently, however, the vine is used for a garden climber and source of the "wood rose" for temperate florist commerce. Even though most of its past is obscure, enough data remain to reconstruct part of its surprisingly complicated history.

ACKNOWLEDGMENTS

This study was supported by a grant from the National Geographic Society (Grant No. 4478-91); publication was supported by the Florida Atlantic University Foundation. Assistance, comments, criticisms, and literature were provided by Eckert Eich (Fachbereich Pharmazie, Institut für Pharmazie II, Freie Universität Berlin, Berlin, Germany), Scott Mori (New York Botanical Garden), George Staples (Bishop Museum, Honolulu), William R. Sykes (Manaaki Whenua Landcare Research, Lincoln, New Zealand), Nicholas Turland (British Museum, London), and the library personnel at Missouri Botanical Garden and FAU. Santiago F. Pérez (FAU) kindly provided the Spanish translation. My wife Sandra helped gather field and herbarium data and provided comments on drafts of the manuscript.

LITERATURE CITED

Adams, C. D. 1972. Flowering plants of Jamaica. University of the West Indies, Mona.

Aiton, W. 1810. Hortus Kewensis; or, A catalogue of the plants cultivated in the Royal Botanic Garden at Kew. ed. 2. 1:339. Printed for Longman, Hurst, Rees, Orme, and Brown, London.

Alcorn, J. B. 1984. Huastec Mayan ethnobotany. University of Texas Press, Austin.

Anonymous. 1807. Transactions of the Horticultural Society of London 1:184. t. 11.

Austin, D. F. 1975. Convolvulaceae. *In* Flora of Panama. Ann. Missouri Botanical Gard. 62:147–224.

———. 1980. Convolvulaceae in Dassanayake. *In* M. D. and F. D. Fosberg, eds., A revised handbook of the flora of Ceylon. 1:288–363. Amerind Publishing Co., New Delhi.

———. 1982a. Convolvulaceae. *In* Z. L. de Febres and J. A. Steyermark, eds., Flora de Venezuela. Caracas, Instituto Nacional de Parques, Ediciones Funcación Educatión Ambiental. Vol. 8, Part 3:15–239.

———. 1982b. Convolvulaceae. *In* Harling, G. and B. Sparre, eds., Flora of Ecuador. No. 15:1–99. Swedish Research Councils, Stockholm.

———. 1990. Convolvulaceae. Pages 548–564 *in* W. L. Wagner, D. R. Herbst, and S. R. Sohmer, eds., Manual of the flowering plants of Hawai'i. University Hawaii Press and Bishop Museum Press, Honolulu.

———. 1997. Parallel and convergent evolution in the Convolvulaceae. Published on WWW at http://www.fau.edu/divdept/biology/people/evolution.htm

———. 1998. Invasive exotic climbers in Florida: Biogeography, ecology and problems. Florida Scientist 61(2):106–117.

———, **and S. Ghazanfar.** 1979. Convolvulaceae, Pages 1–64 *in* E. Nasir and S. I. Ali, eds., Flora of West Pakistan. Agricultural Research Council, Islamabad.

———, **and G. W. Staples.** 1980. *Xenostegia,* a new genus of Convolvulaceae. Brittonia 32(4):533–536.

———, **and P. B. Cavalcante.** 1982. Convolvuláceas da Amazônia. Museu Paraense Emílio Goeldi, Belém.

Backer, C. A. 1965. Flora of Java. Vol. 2:483–498. Noordhoff-Groningen, Netherlands.

Bailey, L. H. 1935. The Standard Cyclopedia of Horticulture. The Macmillan Company, New York.

———. 1949. Manual of cultivated plants. Macmillan Company, New York.

Baker, J. G. 1877. Flora of Mauritius and the Seychelles. Reeve and Co., London.

Blanco, M. 1837. Flora de Filipinas. Segun el sistema sexual de Linneo. C. Lopez, Manila.

Bojer, W. 1837. Hortus Mauritianus: ou Énumération des plantes, exotiques et indigénes, qui croissent a l'Ile Maurice, disposées d'aprés la méthode naturelle. Imprimerie d'Aimé Mamarot et Compagnie, Maurice.

Brako, L., and J. L. Zarucchi. 1993. Catalogue of the flowering plants and gymnosperms of Peru/Catálogo de las angiospermas y gimnospermas del Perú. Monographs in Systematic Botany Vol. 45. Missouri Botanical Garden, St. Louis.

Browne, P. 1756. The civil and natural history of Jamaica. Printed for the author, London. Reprinted by Arno Press, New York. 1972.

Chapman, A. W. 1897. Flora of the Southern United States. Cambridge Botanical Supply Co., Cambridge, MA.

Chang, Ching-en. 1978. Convolvulaceae. *In* Flora of Taiwan. 4:347–389. Epoch Publishing Co., Taipei [edited by Editorial Committee of the Flora of Taiwan; Hui-lin Li, chairman].

Choisy, J. D. 1845. Convolvulaceae. *In* A. DeCandolle, ed., Prodromus. 9:323–465. Geneva.

Clarke, C. B. 1883. Convolvulaceae. *In* J. D. Hooker, ed., The flora of British India. 4:179–228. L. Reeve and Co., London.

Corominas, J. c.1954. Diccionario critico etimologico de la lengua castellana. Gredos, Madrid.

Craib, W. G., and A. F. G. Kerr. 1954. Florae Siamensis enumeratio: a list of the plants known from Siam with records of their occurrence. Siam Society, Bangkok.

Edwards, S. 1823. The botanical register: consisting of coloured figures of exotic plants cultivated in British gardens, with their history and mode of treatment. 9:t. 768. Printed for James Ridgway, London.

Fang, Rhui-cheng, and G. W. Staples. 1995. Convolvulaceae. *In* Wu, Zheng-yi and P. H. Raven, eds., Flora of China. Gentianaceae through Boraginaceae. Vol. 16:271–325.

Farnsworth, N. R., R. N. Blomster, M. W. Quimby, J. W. Schermerhorn. 1969. The Lynn Index. A bibliography of Phytochemistry. Published by N. R. Farnsworth, Pittsburgh, PA.

Ferguson, I. K., B. Verdcourt, and M. M. Poole. 1977. Pollen morphology in the genera *Merremia* and *Operculina* (Convolvulaceae) and its taxonomic significance. Kew Bulletin 31:763–773, pl. 30–33.

Fournet, J. 1978. Flore illustreé des phanerogames de Guadelupe et de Martinique. Institut National de la Recherche Agronomique, Paris.

Gagnepain, F., and P. Courchet. 1915. Convolvulaceae. Pages 228–313 *in* H. Lecompte, ed., Flore Generale de Indo-Chine Vol. 4. Masson et Cie., Éditeurs, Paris.

Garibay K., A. M., ed. 1985. Historia General de las cosas de Nueva España escrita por Fr. Bernandino de Sahagún. ed. 6. Editorial Porrüa, D.F., México.

Gonçalves, M. L. 1987. Convolvulaceae and Cuscutaceae. Pages 1–142 *in* Flora Zambesiaca. Vol. 8, Part 1. Flora Zambesiaca Managing Committee, London.

Gordon, D. R., and K. P. Thomas. 1994. Introduction pathways for invasive non-indigenous plant species. Pages 29–43 *in* D. C. Schmitz and T. C. Brown, eds., An assessment of invasive non-indigenous species in Florida's public lands. DEP, Bureau of Aquatic Plant Management, Technical Report No. TSS-94-100, Tallahassee.

Gray, A. 1878. A synoptical flora of North America. American Book Co., New York.

———. 1886. Synoptic Flora of North America. ed. 2. Ivison, Blakeman, Taylor and Co., New York.

Grisebach, A. H. R. 1864. Flora of the British West Indian Islands. L. Reeve and Co., London.

Gunn, C. R. 1977. *Merremia discoidesperma* (Donn. Sm.) O'Donell: its taxonomy and capacity of its seeds to drift in ocean currents. Economic Botany 31:237–252.

Hegnauer, R. 1964. Convolvulaceae. Pages 547–561 *in* Chemotaxonomie der Pflanzen: eine Übersicht über die Verbreitung und die systematische Bedeutung der Pflanzenstoffe. Band 3. Birkhauser, Basel.

Hallier, H. 1894. Convolvulaceae africanae. Botanische Jahrbücher Systematik 18:159–160.

———. 1897. Bausteine zu einer Monographie der Convolvulaceen. 9. Die von Caec. und Ed. Seler in Guatemala gesammelten Convolvulaceen des Berliner Herbars. Bulletin de l'Herbier Boissier 6:408–418.

Hemsley, W. B. 1882. Biologia Centrali-Americana; or contributions to the knowledge of the fauna and flora of Mexico and Central America. R. H. Porter and Dulau and Co., London.

———. 1892. A drift-seed (*Ipomoea tuberosa* L.). Annals of Botany 6:369–372.

Humboldt, A. von. A. Bonpland, and K. S. Kunth. 1819. Convolvulaceae. Pages 94–124 *in* Nova Genera et Species Plantarum. Reprinted 1963 by Hafner Publishing Co., Inc., New York.

Kahn, M. S. 1985. Convolvulaceae of Bangladesh. Pages 1–59 *in* Salar Kahn, ed., Flora of Bangladesh. Bangladesh Agricultural Research Council, Dacca, India.

LeCointe, P. 1947. Amazônia Brasileira III. Árbores e Plantas Úteis. Companhia Editora Nacional, São Paulo.

Lejoly, J., and S. Lisowski. 1992. Le genres *Merremia* et *Ipomoea* (Convolvulaceae) dans la flore d'Afrique centrale (Zaïre, Ruwanda, Burundi). Fragmenta Floristica et Geobotanica 37(1):21–125.

Leon, H., and H. Alain. 1974. Flora de Cuba, Vol. 2. Otto Koeltz Science Publishers, Koenigstein.

Linajes, A., V. Rico-Gray, and G. Carrión. 1992. Traditional production system of the root of jalapa, *Ipomoea purga* (Convolvulaceae), in central Veracruz, Mexico. Economic Botany 48:84–89.

Linnaeus, C. 1748. Hortus Upsaliensis, exhibens plantas exoticas, Horto Upsaliensis Academiae a sese illatas, ab anno 1742, in annum 1748, additis differentiis, synonymis, habitationibus, hospitiis, rariorumque descriptionibus, in gratiam studiosae juventutis. Sumtu and Literis L. Salvii, Stockholmiae.

———. 1753. Species plantarum, exhibentes plantas rite cognitas, ad genera relatas, cum differentiis specificis, nominibus trivialibus, synonymis selectis, locis natalibus, secundum systema sexuale digestas. Impensis L. Salvii, Holmiae.

Liogier, H. A., and L. F. Martorell. 1982. Flora of Puerto Rico and adjacent islands: a systematic synopsis. Editorial de la Universidad, Río Piedras.

Maddison, W. P., and D. R. Maddison. 1992. MacClade Version 3. Analysis of phylogeny and character evolution. Sinauer Associates, Inc., Sunderland, MA.

McDonald, A. 1982. Biosystematics of the *Ipomoea tricolor* complex (Convolvulaceae). Ph.D. dissertation, University of Texas, Austin.

———. 1987. Three new species of Convolvulaceae from northeast Mexico. Brittonia 39:106–111.

———. 1993. Flora de Veracruz. Convolvulaceae I. Xalapa, Veracruz, Instituto de Ecologica, A. C. and University of California, Riverside.

Meeuse, A. D. J. 1957. The South African Convolvulaceae. Bothalia 6:641–792.

Meisner, C. F. 1869. Convolvulaceae. *In* K. F. P. von Martius, ed. Flora Brasiliensis. 7:200–424. Apud R. Oldenbourg in comm., Monachii, Lipsiae.

Miller, Philip. 1768. The gardeners dictionary. Printed for the author and sold by John and Francis Rivington [and others], London.

Miquel, F. A. W. 1857. Flora van Nederlandsch Indie. C. G. van der Post, Amsterdam.

Monardes, N. 1569. Historia medicinal de las cosas que se traen de nuestras Indias occidentales que se sirven en medicina. Sevilla, Spain.

Mors, W. B., and C. T. Rizzini. 1966. Useful plants of Brazil. Holden-Day Inc., San Francisco, CA.

Nair, G. G., M. Daniel, and S. D. Sabuis. 1984. Chemosystematics of *Ipomoea* L. and some related taxa. Current Science 55:961–965.

OED. 1971. The compact edition of the Oxford English Dictionary. 2 vols. Oxford University Press, Oxford.

O'Donell, C. A. 1941. Revision de las especies americanas de *Merremia* (Convolvulaceae). Lilloa 6: 467–554.

Ooststroom, S. J. 1953. Convolvulaceae. *In* C. G. G. J. van Steenis, ed., Flora Malesiana ser. 1, 4(4): 388–512. Noordhoof-Kolff n.v., Djakarta.

Oviedo, G. F. de. 1526. De la Natural Hystoria de las Indias. Facsimile edition, No. 85, 1969, University North Carolina Press, Chapel Hill.

Plukenet, L. 1691. Phytographia, sive, Stirpium illustriorum & minus cognitarum icones tabulis aeneis: summa diligentia elaborate. Sumptibus Autoris, Londini.

———. 1696. Almagestum botanicum, sive, Phytographiae Pluc'netianae onomasticon methodo synthetica digestum: exhibens stirpium exoticarum rariorum, novarumque nomina, quae descriptionis locum supplere possunt. Sumptibus autoris, Londini.

Powell, D. A. 1977. The voyage of the plant nursery, H. M. S. *Providence,* 1791–1793. Economic Botany 31:387–431.

Powell, D. A., and G. W. Staples. 1989. Convolvulaceae. Pages 133–185 *in* R. A. Howard, ed., Flora of the Lesser Antilles. Arnold Arboretum, Harvard University. Jamaica Plain, MA.

Rani, N., and K. M. Matthew. 1983. Convolvulaceae of the Tamilnadu Carnatic. Pages 1004–1043 *in* The flora of the Tamilnadu Carnatic. St. Joseph's College, Tiruchirapalli, India.

Rendle, A. B. 1905. Convolvulaceae. *In* W. T. Thiselton-Dyer ed., Flora of tropical Africa 4(2):62–206. Kent, England, L. Reeve and Co.; reprint 1959, Beadley Brothers Ltd., Kent.

Roig y Mesa, J. T. 1945. Plantas Medicinales Aromáticas o Venenosas de Cuba. Ministerio de Agricultura, Habana.

Saldanha, C. J., and D. H. Nicolson, eds. 1976. Flora of Hassan District, Karnataka, India. Amerind Publishing Co., New Delhi.

Schimming, T., A. Bufacchi-Richter, K. Jenett-Siems, P. Mann, B. Tofern, B. Dräger, M. P. Gupta, M. D. Correa A., and E. Eich. 1996. Wide distribution of glycosidase inhibiting calystegines in the Convolvulaceae. Paper given at the International Society of Chemical Ecology (ISCE). 13th Annual Meeting, Prague, The Czech Republic.

Schultes, R. E. 1941. A contribution to our knowledge of *Rivea corymbosa,* the narcotic ololiuhqui of the Aztecs. Harvard Botanical Museum, Cambridge, MA.

———**, and A. Hofmann.** 1979. Plants of the gods. McGraw-Hill Book Co., New York.

Shellard, E. J. 1961. The chemistry of some convolvulaceous resins Part II. Brazilian jalap. Planta Medica 9:141–145, Part III 9:146–152.

Simpson, C. T. 1916. Ornamental gardening in Florida. Publ. by the author, Little River [Miami, FL].

Sloane, H. 1696. Catalogus plantarum qua in insula Jamaica. Published by the author, London. Catalogus plantarum quae in insula Jamaica sponte proveniunt, vel vulgò coluntur, cum earundem synonymis and locis natalibus: adjectis aliis quibusdam quae in insulis Maderae, Barbados, Nieves, et Sancti Christophori nascuntur. Seu Prodromi historiae naturalis Jamaicae pars prima. D. Brown, Londini.

———. 1707. A voyage to the islands Madera, Barbados, Nieves, S. Christophers, and Jamaica with the natural history of the herbs and trees, four-footed beasts, fishes, birds, insects, reptiles, etc. 1:1–264. Published by the author, London.

Small, J. K. 1913. Flora of Miami. Published by the author, New York.

———. 1933. Manual of the Southeastern Flora. University North Carolina Press, Chapel Hill.

Standley, P. C. 1938. Convolvulaceae, in Flora of Costa Rica. Field Museum of Natural History, Botany Series 18:960–974.

Swofford, D. L., and D. P. Begle. 1993. Phylogenetic analysis using parsimony (PAUP). Version 3.1.1. Illinois Natural History Survey, Champaign, IL.

Teran, A. L., and J. M. Kingsolver. 1977. Revision del genero *Megacerus* (Coleoptera: Bruchidae). Opera Lilloa 25:1–287.

Verdcourt, B. 1963. Convolvulaceae. Pages 1–161 *in* Flora of tropical East Africa. Crown Agents for Oversea Governments and Administrations, London.

Walker, E. H. 1976. Flora of Okinawa and the Southern Ryukyu Islands. Smithsonian Institution Press, Washington, DC.

Welwitsch, F. 1858. Apontamentos phyto-geographicos: sobre a flora da provincia de Angola na Africa equinocial servindo de relatorio preliminar cerca da explorasão botanica da mesma provincia executada por ordem de sua magesta de fidelissima pelo Dout Frederico Welwitsch. pp. 527–593. Annaes do conselho ultramarino, Lisboa.

Williams, L. O. 1970a. An antivenin, a pacifier and a bit of botanical sleuthing. Economic Botany 27: 147–150.

———. 1970b. Convolvulaceae. *In* P. C.Standley and L. O. Williams, eds., Flora of Guatemala. Fieldiana: Bot. 24(Part IX):4–85.

———. 1970c. Jalap or Veracruz Jalap and its allies. Economic Botany 24:399–401.

Ximénez, F. 1615. Plantas, Animales, y Minerales de Nueva España, Usados en la Medicina. *In* M. Leon, ed., Reprinted 1888, Governo de Oaxaca, Mexico.

Yuncker, T. G. 1943. The flora of Niue Island. Bulletin Bernice P. Bishop Museum 178.

Notes on Economic Plants

Pandanus, Screwpine Painting in Fifth Century Buddhist Caves at Bagh, Madhya Pradesh, India.—Bagh Caves in Dhar district, Madhya Pradesh (22°22′N and 77°78′E) were excavated and built by one Dattaka, a Buddhist Bhikku, for the residence of Buddhist monks from all over India (*1*). A copper plate inscription found in Cave 2 further records that one king Subandhu in the fifth century AD donated eight villages for the upkeep and maintenance of the Caves. Almost the whole group of nine caves has now crumbled not only because of the weak sandstone but also because of the thoughtless removal in the 1950s of tree and shrub vegetation of the hillock which had earlier been a factor for stability.

An ambitious restoration work on the caves, according to Singh (*2*) was begun in 1982. A bridge on the river Baghini has been built for smooth and all-year-round access to the caves. The technical details of restoration work of the Bagh Cave paintings as of other caves in India have been reviewed admirably by Subbaraman (*3*). Because of the process of clean-up and removal of grime, soot and other kinds of dirt and debris, some new paintings have been uncovered. Although most are rather now faint or altogether obliterated, their occasional brilliance is quite astonishing. It seems the colors have remained more or less in the original state because of the negligible exposure to light.

Attention had earlier been drawn to Mushroom paintings in the ceiling of Cave 4 by Payak (*4*). This cave was known as Ranga Mahal (palace of colors) or as Kalayan (Art abode) as stated by Sankalia (*1*) in the newspaper article. Subbaraman (*3*) has indicated that the paintings on the walls as well as on the ceiling, as part of the restoration work, have been detached section by section and will be fixed after proper treatment.

During a visit in February 1990, I was shown some wall paintings in the first cave from the left that came to light after removal of dirt, soot, etc. The paintings are on the right side near the innermost corner and can be viewed only with the help of artificial illumination. Among these, the painting of three male spadices of pandanus was of interest and is the subject of this communication.

Pandanus odoratissimus L.f. featured prominently in religious, social and linguistic contexts since ancient times. The plant under the name of *ketki* has been extolled in many Sanskrit writings, Puranic and Classical. In Agni Puran, one of the 18 purans, the Agni god prescribes the various types of flowers to be used in the worship of god Vishnu. Among these the leaves and flowers of *ketki* also find a mention; *ketki* is also cited in many *shloks* (or stanzas) of Valmiki's Ramayana. But its association with Kamadeva (the god of Erotics) together with many images in sensual contexts comes out vividly in the *Ritusamhar* of Kalidas, one of the earliest and greatest poets of classical Sanskrit during the Golden Age of India. *Ritusamhar* quoted here is that edited and provided with Hindi translation by Sharma (*5*). The poet delineates with great delicacy the delights and pleasures of all the six seasons of the year. The flowering of *ketki* along with three other plants, represents not just the onset but the very acme and the ambience of the rainy season (*Varsha Ritu*). In Sarga (chapter) 2 which follows the first one on the "Hot" or the summer season (*Grishm Ritu*), *ketki* figures in four *shloks* or stanzas. In *shlok* 17, Kalidas talks of a breeze of moist air which sends shivers through the forests of *ketki* and Sal (*Shorea robusta* Gaertn. F.); in 20, he states that ladies wear on their heads garlands woven out of *ketki* flowers; in 23, it is stated that with the advent of the new season the forest country is smiling with the buds of *ketki*; and finally in *shlok* 26, again it is the breeze which sweetly/deliciously troubles the inhabitants with the welcome fragrance of *ketki raj* (powder or pollen?). Thus from very early times, possibly two thousand years or more, the flowers of *ketki* or pandanus have been used in worship, as an ornament for ladies and as a source of fragrance or perfume in India.

The question arises how the flowers of such a plant should have been painted where asceticism and celibacy had to be strictly observed? Undoubtedly, Lord Buddha had expressly pro-

Economic Botany 52(4) pp. 423–425. 1998

hibited many trades like painting and certainly images of human forms in any way whatsoever were not to be depicted or painted especially in monasteries or Vihars where a large congregation of monks or Bhikshus lived. However, especially after Ashok the Great, two hundred years or so after the Mahaparinirvana or the Great Decease of Lord Buddha, this prohibition possibly was dispensed with altogether. On the other hand, painting or sculpture came to be regarded as a vehicle for the pious purpose, of spreading Buddhism; this was particularly the case with the followers of the Mahayana sect. Moreover, Lord Buddha, according to Dhanpala (*6*), was not against painting of plants, flowers, birds or other objects found in Nature.

Randhawa (*7*) has stated that in Mughal times, the plant received a good deal of attention at the hands of emperors Babur, Akbar and Jahangir. Two types of pandanus were mentioned in Ain-i-Akbari, composed and chronicled by the famous courtier of Akbar, Abu-l-Fazl; Jahangir also confirmed that there were two kinds of pandanus, *ketki* and *kewra.* The former was stated to have the form of a spindle with 12 or more petals and emitted a delicate scent; *kewra* was stated to be similar but was twice bigger than *ketki.*

Three perfumery products, known as *kewda attar, kewda water and kewda oil* (*rooh kewda*), are obtained from a variable species, *Pandanus odoratissimus.* In the present times, as stated in (*8*) and also by Datta et al (*9*), the perfume industry based on this plant is mainly confined to the district of Ganjam in Orissa; on a minor scale, however, the perfume is extracted at a few places in the states of Andhra Pradesh, Tamil Nadu and Uttar Pradesh also.

According to Santapau and Henry (*10*), about 36 species are known to occur in India. *Pandanus odoratissimus* includes an unarmed form, forma *laevis* which is cultivated for extracting the perfume; forma *samek* includes prickly types the tough leaves of which are used for making mats, nets, cordage, sacks, etc; lastly, forma *variegatus* is grown as an ornamental because of the longitudinal bands that the leaves display. With all these characters, the overall appearance of the plant is very ungainly and asymmetric. It occurs in moist situations, along the river banks, canals, ponds, wells, etc. throughout India. Because of the general apathy towards things traditional including perfumes such as attars and scents, even among floriculturists and conservationists of indigenous germplasm, there is very little interest in its cultivation and survival. The plant is much less common now than it was a few decades ago.

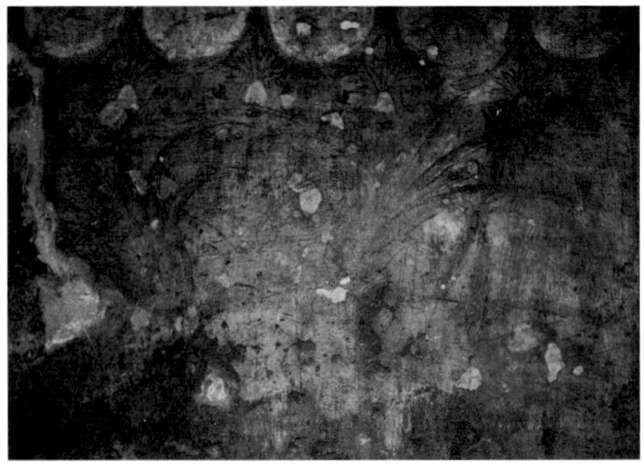

Fig. 1. Wall painting in Cave 1 at Bagh. Note three male spadices of pandanus.

The fragrant smell is emitted by the tender spathes or bracts covering the flowers of the male inflorescence. They are white to yellow inside and green outside and armed with spines on the margins in certain forms as stated above. The spadices depicted at Bagh (Fig. 1) represent a prickly type of pandanus. Undoubtedly, the plant now is rather uncommon in Dhar district of Madhya Pradesh in which the Bagh Caves are situated. But the author has observed a few colonies of the prickly types near temples and wells in the town of Dhar. Its occurrence around Bagh Caves situated on the banks of the river Baghini, at the time the caves were excavated and painted thus can not be ruled out.

Literature Cited. (*1*) Sankalia, H. D. 1985. Crumbling Bagh Caves Hitwada. (February 17 Issue): 4–5. (*2*) Singh, N. K. 1983. A wasted heritage. Indian Express (December 25 Issue). (*3*) Subbaraman, S. 1993. Conservation of mural paintings, Current Science. 64:736–752. (*4*) Payak, M. M. 1984. Mushroom paintings in Buddhist Caves of Bagh. Roopa-Lekha. 55:41–47. (*5*) Sharma, K. ed. 1962. RituSamharam of Kalidas (Prabha Commentary)—In Sanskrit with Hindi translation. Chaukhamba Vidya Bhawan, Varanasi; (*6*) Dhanapala, D. B. 1964. Buddhist paintings from Shrines and Temples in Ceylon. Mentor-UNESCO, New American Library. New York. (*7*) Randhawa, M. S. 1982. A history of agriculture in India, Vol.II. The Indian Council of Agricultural Research, New Delhi; (*8*) Wealth of India. 1987. Vol.11 (N to Pe). The Council of Scientific and Industrial Research,

New Delhi; (*9*) Datta. P. K., H. O. Saxena, and M. Brahman. 1982. Kewda distillation Industry in Orissa, pp. 634–639 in C. K. Atal and B. M. Kapur. eds., Cultivation and utilization of aromatic plants. Regional Research Centre, The Council of Scientific and Industrial Research, Jammu; (*10*) Santapau, H. and A. N. Henry. 1973. A dictionary of the flowering plants of India. The Council of Scientific and Industrial Research, New Delhi.

—M. M. Payak, Pusa Apartments, Sector 15, Rohini, New Delhi-110085, India.

BOOK REVIEW

Sacred Leaves of Candomblé. African Magic, Medicine, and Religion in Brazil. Robert A. Voeks. 1997. University of Texas Press, P.O. Box 7819, Austin, TX 78713-7819. xx + 236 pp. (paperback). $17.95 (paperback), $37.50 (hardcover). ISBN 0-292-78731-6 (paperback); 0-292-78730-8 (hardcover).

The topic covered in this book is the ethnobotany of the Afro-Brazilian religion Candomblé. The author begins his extraordinary analysis of plant use by Afro-Brazilians by describing the geographic and climatic factors (starting in the Cretaceous!) that have shaped the physical characteristics of both eastern Brazil and western Africa. He points out that, although biological factors created a Brazilian landscape that was significantly different from west Africa, farming, forestry, and animal husbandry created a landscape dominated by sugar, cocoa, oil palm, coffee, and pasture grasses recognized by African slaves who were forced across the Atlantic. Voeks clearly explains the evolution of plant use in this religion and states that "Afro-Brazilian ethnobotany would never be based in knowledge of the jungle. Their understanding of plants . . . would be assimilated from indigenous sources, picked up from their European masters, and ultimately, blended with the traditional knowledge of their ancestors." Colonial medicine, indigenous and slave medicine and magic are then discussed as they pertain to and impact treatment and disease conceptions amongst the Afro-Brazilians. Brazilian Candomblé is defined and explained in great detail with a wonderful discussion of the evolution of this very complex religion. The very complicated healing practices of this religion in reference to the use of plants for both ritual and medicinal purposes are explained; an appendix which provides detailed ethnobotanical information completes Voek's discussion.

I have not seen such a thorough ethnobotanical study of any of the West African influenced New World religions. The author spends time not only describing the ethnobotany of this religion as it exists today but also examines the evolution of this religion and its plant lore. Voeks eloquently states,

> ". . . this book is not about the highly touted medicinal potential of pristine tropical rainforests. Rather, it underscores the intrinsic medicinal worth of peopled tropical landscapes, of disturbed forests and fields and the healing flora they harbor. The sacred leaves of Candomblé are, with few exceptions, sacred shrubs, sacred weeds, and sacred cultivars. Depending on your perspective, this feature either undermines any botanical interest in the Afro-Brazilian pharmacopoeia or, on the contrary, suggests that a significant portion of the 'tropical pharmaceutical factory' should be sought in disturbed as opposed to primary forests."

This is a bold statement with many implications. The *Sacred Leaves of Brazilian Candomblé* really made me stop and think about the ethnobotanical importance of human-derived landscapes and the study of such ecosystems. I highly recommend this book for all ethnobotanists, particularly those interested in African American ethnobiology.

KRISTEN PORTER-UTLEY
DEPARTMENT OF BOTANY
3165 MCCARTY HALL A
UNIVERSITY OF FLORIDA
GAINESVILLE, FL 32611-8526

Announcement

The Anne S. Chatham Fellowship in Medicinal Botany

The Garden Club of America is pleased to announce a new grant open to Ph.D. candidates and Ph.D.s to enable study in medicinal botany. The fellowship was established to protect and preserve knowledge about the medicinal use of plants, and this prevent the disappearance of plants with therapeutic potential. Providing this research opportunity for botanists can, in turn, assist medical science in its ability to protect lives and develop new medicines. The fellowship was created and endowed by a generous contribution from Anne S. Chatham, a member of the Garden Club of America's Scholarship Committee and a member of the Jupiter Island Garden Club.

To this end, the Anne S. Chatham Fellowship in Medicinal Botany will provide one grant award of $4000 annually. The Missouri Botanical Garden has agreed to administer the new fellowship and selection of candidates for the award will be by a panel of botanists specializing in Economic and Ethnobotany. Eligible applicants should send a brief application letter, 1–2 page description of the proposed research, and current curriculum vitae to Dr. James S. Miller, Missouri Botanical Garden, P.O. Box 299, St. Louis, MO 63166-0299 by 15 January 1999. Announcement of the successful candidate is expected in March 1999. Additional information can be obtained at the Anne S. Chatham Fellowship site on the World Wide Web at: http://www.mobot.org/mobot/research/applied_research/chatham.html or by contacting Dr. James S. Miller at the address above, by phone (314) 577-9503, or email at miller@mobot.org.

Announcement

New Crops and New Uses: Biodiversity and Agricultural Sustainability. Hyatt Regency at Civic Center, Phoenix, Arizona November 8–11, 1998

Sponsoring organizations: The Association for the Advancement of Industrial Crops (AAIC), Perdue University Center for New Crops and Plant Products, and The New Uses Council, Inc.

Session topics will include: World Biodiversity Update, Strategic Opportunities of New Crops and New Uses, Biodiversity and Industry Issues, Processes and Strategies in Commercializing New Crops and New Uses.

Panel discussions and special topics will be hosted by AAIC divisions. A poster session is scheduled with awards presented for outstanding and student presentations. Exhibit booths will be available for displays.

Poster presentation manuscripts may be submitted for publication in *New Crops and New Uses: Biodiversity and Agricultural Sustainability,* a book to be published after the symposium and included in the registration fee.

For further information or to pre-register, please contact: David A. Dierig, President AAIC % US Water Conservation Laboratory 4331 E. Broadway Rd. Phoenix, AZ 84050; Telephone: (602) 379-4356; Fax: (602) 379-4355; Email: ddierig@uswcl.ars.ag.gov.

Economic Botany 52(4) pp. 426. 1998

Corrections

Useful Plants of the Las Tuxlas Rain Forest (Veracruz, Mexico): Considerations on their Market Potential *Guillermo Ibarra-Manríquez, Martin Ricker, Guillermo Angeles, Santiago Sinaca Colín, and Miguel Angel Sinaca Colín* 51(4)362–376. 1997

p. 372 *Desmoncus orthacanos,* common name is "junco negro."
p. 373 in Clusiaceae, correct names are: *Calophyllum brasiliense* Cambess. Triana & Planch.; and *Rheedia edulis* (Seem.).
Diospyros digyna Jacq./*GIM3495*/Zapote negro, Zapote prieto/M,C/Tree/Edible fruits (Sep–Jan) [correct sequence is L1,N,I. (2, 16, 17, 21, 22)]
Common name for *Sloanea medusula* Schum. & Pittier is "Erizo."
p. 374 *Virola guatemalensis,* specimen number is GIM2450.
Erythrina folkersii specimen number is GIM3172.
Piper auritum, is a shrub or tree and leaves are used as a spice.

Book Review

Promoting the Conservation and Use of Underutilized and Neglected Crops. 8. Chayote. *Sechium edule* (Jacq.) Sw. Rafael Lira Saade. 1996. International Plant Genetic Resources Institute (IPGRI), Via delle Sette Chiese 142, 00145 Rome, Italy. 58 pp. (paperback). Gratis. ISBN 92-9043-298-5.

Rafael Lira did his Ph.D. dissertation on this gourd and its relatives. He is uniquely qualified to write what amounts to a small monograph on this one species.

A major problem confronting the breeder who might want to work on improving any crop is determining what the other species are, and which ones it might cross with. It appears from the summary on page 10 that the species are fairly well agreed upon; the issue of whether to divide *Sechium* into as many as three genera is not resolved. (But see Lira, R., V. L. Villaseñor, and P. D. Davila, Systematic Botany 22 (2): 415–426. 1998, where the conclusion to treat *Sechium* as a single, inclusive genus is strongly supported.)

Chayote was cultivated long before the Spaniards came to the New World. From its origins in southern Mexico and Central America, the plant has been spread worldwide. It is reported, Lira tells us, that the plant has even become a weed in Java. In India, chayote finds use not only as a vegetable but also as cattle feed. The major exporter is Costa Rica, but it appears the value of the crop is not that great, varying from U.S. $5–8 million per year.

Clearly, there's a lot of work to be done before chayote will become a significant crop. One wonders whether a clever tradename, like "Kiwi Fruit" for *Actinidia chinensis,* would help to boost sales. The fruit already has a number of local names (p. 13), and perhaps one of those will, with clever advertising, become popular.

The plant has a surprising number of pests (pp. 46–48), which probably do little harm to the plant in nature. Under cultivation, however, the matter is very different, because fruits of export quality have to be unblemished and free of parasites. Lira suggests one way out of this is to promote the use of the fruit pulp in industrial processes: it can be used for giving a pleasing consistency to baby foods, juices, sauces, and pastes (p. 29), and then how pretty the ovary wall is becomes irrelevant.

Like all the books in this series, the English is perfect, the typography and photo reproduction are excellent, and the bibliography is extensive but there's no index.

Neil A. Harriman
Biology Department
University of Wisconsin–Oshkosh
Oshkosh, Wisconsin 54901 USA

Economic Botany 52(4) pp. 427. 1998

BOOK REVIEWS

DANIEL F. AUSTIN, BOOK REVIEW EDITOR

Cucurbits. Crop Production Science in Horticulture 6. R. W. Robinson and D. S. Decker-Walters. 1997. CAB International, distributed by Oxford University Press, 198 Madison Avenue, New York, NY 10016. viii + 226 pp. (paperback). $40.00. ISBN 0-85199-133-5.

"Cucurbits are weeds waiting to become crops" begins a chapter on crop evolution and plant use. Admittedly over-simplifying evolution in the Cucurbitaceae, the authors note that the often large or showy fruits of many cucurbits probably attracted the attention of aboriginal gatherers. That attraction eventually lead to some of the earliest domesticated plants. The economic value and taxonomic diversity used by humans is evident in the book's first chapter, "What Are Cucurbits?" Table 1.1 lists 28 genera and 48 species of cultivated cucurbits grouped according to food, medicinal, ornamental, or utilitarian value. The four major cucurbit crops (watermelon, cucumber, melon, and squash) and several of the five other common ones (bitter melon, bottle gourd, chayote, loofah, wax gourd) are produced in increasing quantities worldwide. This increase is due to use of greater amounts of land, improved yields, or enhanced specialty markets.

This book is part of a series on temperate, subtropical and tropical horticultural crops from CAB International. The 16 books emphasize "scientific principles underlying crop production practices rather than . . . providing empirical recipes for uncritical acceptance." The series is primarily aimed at university and college students and staff, but according to the editors, would be informative for "progressive" growers, "keen" gardeners, and production consultants.

The co-authors of this volume complement each other. Robinson is a plant breeder and Decker-Walters is a systematist. Both have extensive knowledge about evolution and management in a broad array of cucurbit resources. The eight chapters range from botany (taxonomy, morphology, anatomy, physiology, evolution, utilization) and genetics (cytology, genes, breeding, genetic engineering) to horticulture (cultural requirements, fruit and seed production), and pest problems (diseases, nematodes, insects, and spider mites). High quality photographs, tables, and graphs—some reprinted, some new—illustrate the text. The book serves as an update to the 1962 classic, *Cucurbits: Botany, Cultivation, and Utilization* (Interscience Publishers, NY) by T. W. Whitaker and G. N. Davis.

Information about the nine most widely grown crops accounts for much of the book. However, descriptions are provided for numerous minor crops. Chapters 2 (Evolution and Exploitation) and 4 (Major and Minor Crops) are of greatest interest to economic botanists. In those chapters, botanical descriptions and information on origin, distribution, and cultural practice are provided for more than 30 species. An appendix lists nearly 80 common names (only a few are non-English terms) and over 50 scientific name equivalents, helpfully sorted first by common name, and then by scientific name. The bibliography is "selected," but lengthy (more than 175 cited publications).

The book contains an error regarding a set of African watermelon relatives (*Citrullus*). In West Africa, two types of cucurbits used mainly for seed consumption are known locally as "egusi". The authors correctly state that one of those is *Cucumeropsis mannii* (also known as the white-seeded melon). But the book mistakenly refers to two watermelon taxa—*Citrullus lanatus* var. *citroides* ("citron" or "tsama") and *C. colocynthis* ("colocynth")—as "egusi", neither of which are known by that name. Instead, the second West African "egusi" is a distinctive landrace of *C. lanatus* var. *lanatus*, the same botanical variety as most of the world's cultivated varieties of watermelon. A partial explanation for the confusion is that African publications typically refer to "egusi" as *Colocynthis citrullus* (a synonym for *C. lanatus*). Then, this is mistakenly assumed to be a reference to the "colocynth." The "citron," on the other hand, is the likely progenitor of the watermelon lineage and is distributed throughout southern Africa. On that continent it serves as virtually the only water source for much of the year to the Kalahari Bushman and other desert inhabitants.

LAURA C. MERRICK
SUSTAINABLE AGRICULTURE PROGRAM
DEPT. OF APPLIED ECOLOGY &
ENVIRONMENTAL SCIENCES
UNIVERSITY OF MAINE
5722 DEERING HALL
ORONO, ME 04469-5722

Cycles of Life. Civilization and the Biosphere. Vaclav Smil. 1997. Scientific American Library, a division of HPHLP; distrubuted by W. H. Freeman & Co., Publishers, 41 Madison Avenue, New York, NY 10010. xii + 221 pp. (hardcover). $32.95. ISBN 0-7167-5079-1.

First off: for whom is this book intended? This is certainly not an introductory textbook—it has none of the apparatus of those, like chapter summaries and

Economic Botany 52(4) pp. 428–440. 1998

"critical thinking" exercises at the end of each chapter. The book attempts to bring together all the pieces of the carbon cycle, the nitrogen cycle, and the sulfur cycle. But there are no footnotes, and no literature cited. The laudatory blurbs on the back of the dustjacket are all written by people with advanced degrees in atmospheric science and the like. The reader had better be thoroughly grounded in introductory chemistry and introductory biology, at a minimum.

You are expected to know what oxidation and reduction are; net primary productivity may be defined somewhere, but I didn't find it, and these words are not in the index. The relevant equations for all these cycles are given in an appendix, I discovered too late in the process, but in a most unusual fashion, because none say anything about ATP yield. The result is that reactions peculiar to organisms are distinguished from wholly non-biological reactions (like the weathering of calcite) only by the inclusion of an organism's name. The Latin names in the equation appendix are mostly italicized, but not all are—it keeps you on your toes, I'll say that for it.

You may still try to read *Scientific American*. If you do, have you ever noticed that any article outside your own field goes along just fine until the end of the third paragraph, and then you get whomped? That's what this book does.

If you can get past that, and I urge you to do so, there are some total gems incorporated here. For example, sulfur dispersed in the atmosphere by burning fossil fuels is widely claimed to form sulfuric acid, which then kills the Black Forest of Germany. Smil points out at some length that this is a crude oversimplification. Read it before you give your next talk to the Kiwanis at noon.

Claims were once made, most famously by Barry Commoner in the late 1960s, that nitrogen fertilizers added to farm fields disrupted the natural nitrogen cycle. Not quite true. Smil dissects the arguments nicely on p. 131. Global warming associated with increasing levels of atmospheric carbon dioxide receives major attention. The strands of the argument are densely interwoven, and it may take a second reading to get them all settled in one's mind. In general, Smil thinks global warming has been much exaggerated,[1] and he summarizes numerous studies that suggest that higher carbon dioxide levels will increase crop yields dramatically. By the same token, the issue of northward migration of whole biomes is not ignored. One point I haven't seen elsewhere: insect vectors for a variety of most unpleasant diseases (like malaria and dengue fever) would also spread northward and southward in a warmer world. This is emphatically not a book for the general reader. But the effort it takes to read it will be well repaid, and I urge you to do so.

NEIL A. HARRIMAN
BIOLOGY DEPARTMENT
UNIVERSITY OF WISCONSIN–OSHKOSH
OSHKOSH, WI 54901 USA

[1] Editor's Note: For other current alternate views, see: Sagan, C. 1997. Billions and Billions. Random House, New York; Bowen, M. 1998. Natural History 107:28–41; Brecher, Henry H. and Thompson, Lonnie G. 1993. Photogrammetric Engineering & Remote Sensing 59:1017–1022; Gregory, J.M. and Oerlemans, J. 1998. Nature 391: 474; Zuo, Z. and Oerlemans, J. 1997. Climate dynamics 13: 835; Turpin, O.C., Ferguson, R.I. and Clark, C.D. 1997. Physics and chemistry of the earth 22: 279; Singh, P. and Kumar, N. 1997. Journal of hydrology 193:316.

American Herbal Products Association Botanical Safety Handbook. Michael McGuffin, Christopher Hobbs, Roy Upton and Alicia Goldberg (eds.). 1997. CRC Press, 2000 Corporate Blvd., N.W. Boca Raton, FL 33431. 231 pp. + 18 un-numbered (hardcover). $30.95. ISBN 0-8493-1675-8.

The use of herbal products for health and personal care is growing in North America at a rate of almost 20% per year and markets contain an unprecedented range of herbal products from around the world. Lacking are sources of reliable information on the properties of particular plants and credible authorities to provide information in a manner that adequately serves the public interest. More than ever consumers are responsible for informing themselves, and practitioners of both conventional and non-conventional health care need to be better prepared to offer advice that is based on the best available scientific data. In publishing the *Botanical Safety Handbook* the American Herbal Products Association makes a valuable contribution that should enhance its credibility as a body responsible to both the needs of its own professional and industrial members and to the public at large.

The resultant volume is comprised of a series of over 500 short entries on safety arranged by Latin binomials that represent the most important herbs currently available in the USA. Plants are given a safety classification and, where appropriate, additional notes are provided which include references to specific toxicity and pharmacological data, to dose, and to the regulatory status in the USA and Canada. Following are three appendices that summarize data on toxicity of specific toxic plant constituents, on different modes of physiological and toxicological activity, and on the safety classification designations put forward in the volume. Citations are generally made to a bibliography of 30 sources plus supplemental publications, although no citations support the many species that are classified as safe without restrictions. Nonetheless, the resultant volume is a unique and highly useful body of reference material. As well as being comprehensive, it thoroughly covers herbs currently receiving attention related to toxicity as well as the constituents of greatest concern such as pyrrolizidine alkaloids.

The presentation of the book is basically scientific and professional, and is far removed from the pseudo-scientific character of much of the information disseminated by herbal entrepreneurs. However, as the responsible herbal industry strives to establish credibility its strongest detractors come from the medical and scientific communities. It is unfortunate then that this book, contains passages that mostly seem simply scientifically unsophisticated but sometimes represent factual errors. Personally I was drawn to regular, albeit minor errors, in use of the terminology and data related to phytochemistry and biochemistry. Synthesis of a far-reaching body of data such as compiled here is a challenge for anyone working beyond their speciality. The committee of non-scientists who complied this volume have succeeded in this regard no less than most experts. However, such weaknesses only feed suspicion of herbalism. While it is regrettable that more effort was not made to ensure that the volume was adequately reviewed by the appropriate scientific and technical specialists before publication, *the Botanical Safety Handbook* is still highly recommended for anyone with a professional or personal interest in the use of herbal medicine.

TIMOTHY JOHNS
CENTRE FOR INDIGENOUS PEOPLE'S NUTRITION AND THE ENVIRONMENT (CINE)
MACDONALD CAMPUS, MCGILL UNIVERSITY
STE. ANNE DE BELLEVUE, QC H9W 5B6
CANADA

Promoting the Conservation and Use of Underutilized and Neglected Crops. 15. Black Nightshades. *Solanum nigrum* L and Related Species. Jennifer M. Edwards and J. A. Chweya (1997). Rome Italy, International Plant Genetic Resources Institute (IPGRI), Via delle Sette Chiese 142, 00145 Rome, Italy. 113 pp. (paperback). Gratis. ISBN 92-9043-321-3.

The International Plant Genetic Resources Institute has a prestigious history of coordinating and promoting the collection, conservation and use of germplasm from the world's major crop plants. While scientists have lamented the vulnerability of global agricultural productions systems based on a handful of major species, efforts to promote development of alternative plants of high potential have been relatively limited. The indigenous crop plants of various regions have received recognition as having local importance particularly in contributing to dietary balance and food security, and publications of the National Academy of Sciences, FAO and others have made the importance of minor foods plants known to a broader audience of scientists and policy makers. Since 1996 the IPGRI has further advanced the promotion of underutilized and neglected crops through a series of species-based monographs, of which "Black Nightshades. *Solanum nigrum* L. and Related Species" is the fifteenth.

Black nightshades comprise a highly variable species complex of *Solanum* section *Solanum* that center around *Solanum nigrum.* The section shows its greatest diversity in South America but *Solanum nigrum* is Eurasian in origin. While the 13 species presented in this volume are all used through most of their respective ranges in Africa, Asia, Australia, Malesia and the Americas, the ethnobotany of section *Solanum* is confused by the tendency to refer to all members as *S. nigrum.* It is clear, however, that leaves and stems of most of the species are used as potherbs, many have edible fruits and they have widespread medicinal uses. Few of these species are cultivated, with most use being of plants collected from the wild or growing spontaneously in fields and gardens.

This volume deals with this group in several topically arranged chapters dealing with taxonomy, systematics, properties and uses, genetic resources, ecology and agronomy and related subjects. All are thoroughly referenced. As well the book has appendices providing research contacts around the world and addresses of gene banks maintaining collections. The body of data compiled demonstrates that this a potentially valuable crop and the book will be a valuable contribution to efforts to enhance black nightshades as a crop through cultivation and breeding.

TIMOTHY JOHNS
CENTRE FOR INDIGENOUS PEOPLE'S NUTRITION AND THE ENVIRONMENT (CINE)
MACDONALD CAMPUS, MCGILL UNIVERSITY
STE. ANNE DE BELLEVUE, QC H9W 5B6
CANADA

Going Nuts for the Rainforest. Non-Timber Forest Products, Forest Conservation and Sustainability in Amazonia. Willem Assies. 1997. Thela Latin American Series, Thela Publishers, in association with the Tropenbos Foundation, Amsterdam, The Netherlands. 96 pp. (paperback). U.S. $18.50, Dfl. 27.50. 9.95. ISBN 90-5538-027-X.

Going Nuts examines the development of the rubber and Amazon nut industries and their role in tropical forest conservation in northern Bolivia and the adjacent Brazilian state of Acre. This study critically explores the role these industries play in the conservation of Amazonian forests through the creation of extractive reserves.

The book provides thorough analyses of the historical, political, ecological, and economic factors responsible for the development of extractive reserves in this region. The author introduces (Chapter 1) the sustainable development debate using a political & ecological framework. Then (Chapter 2) he describes the historical development of the extractive economy in Acre and northern Bolivia, and the changes these economies have undergone beginning in the middle nineteenth century. In chapter 3, the international and national political conditions are discussed that resulted in the development of the modern extractive reserve concept over the past 25 to 30 years, and how extractive reserves evolved from a land reform measure into a conservation tool. Chapter 4 describes the emergence of northern Bolivia as a competitor in the extractive market place and what this competition means for conservation and the extractive reserve concept in Acre.

Assies describes the inner workings of the extractive reserve economy to support his thesis that their long-term viability is doubtful. For example, the production of Amazon nuts has developed in close association with rubber tapping because their harvest seasons do not overlap. Profits from neither are great enough alone to make extraction focused on a single product a profitable endeavor. As a result, if one economy falters, the entire extractive economy is depressed and extraction of all resources is abandoned by many. According to Assies, this is what happened when the Brazilian government recently decided to abandon price supports for rubber. Many of the rural harvesters abandoned the forest and resettled in urban areas in search of jobs, or expanded their agricultural efforts at the expense of the forest. Thus, much of the remaining extraction of Amazon nuts is carried out not by traditional forest dwelling peoples, but by contract labor hired in regional urban centers.

If the economic viability of extractive reserves is dependent on subsidies, Assies feels their long-term success is doubtful given the current global emphasis on a free market and deregulated economic system. For extractive reserves to serve as a long-term conservation solution, they must be more profitable than other forms of extractive economies, such as logging and cattle ranching. According to the author, they presently fall short of this goal.

One shortcoming of the book is the lack of local peoples' opinions. Although household surveys were alluded to, the author's arguments could have been strengthened by interviews and direct quotes taken from local people involved in extractive activities. Instead, the author relied heavily on economic statistics, which although informative, lacked the anecdotal insights of those involved in rubber and Amazon nut industries.

This book would be valuable for those interested in extractive reserves specifically, but also anyone interested in tropical conservation issues, sustainable development, and the emerging field of political ecology. Assies provides a candid up-to-date analysis of one of the more popular conservation tools developed in recent years. He demonstrates that placing a monetary value on non-timber forest products is relatively simple compared to turning those potential numbers into realized profits for local people.

Michael Kenneth Steinberg
Department of Geography & Anthropology
Louisiana State University
227 Howe/Russell Geoscience Complex
Baton Rouge, LA 70803-4105

Promoting the Conservation and Use of Underutilized and Neglected Crops. 14. Oregano. Proceedings of the IPGRI International Workshop on Oregano, 8–12 May 1996, CIHEAM, Valenzano (Bari), Italy. Padulosi, S. (ed.) 1997. International Plant Genetic Resources Institute (IPGRI), Via delle Sette Chiese 142, 00145 Rome, Italy. 176 pp. (paperback). Gratis. ISBN 92-9043-317-5.

This series typically includes the Latin name of the plant in the title. This one does not, because there are a dozen species of *Origanum* (Lamiaceae) alone that may comprise what we call oregano on the spice shelf. In addition, there are various other genera of Lamiaceae, such as *Hyptis, Coleus* (now to be submerged into *Plectranthus*, but never mind), *Monarda, Calamintha, Salvia,* and *Thymus*. One can also make oregano from *Lippia* and *Lantana* (Verbenaceae). What they all have in common is they produce thymol and carvacrol, the oils that create the distinctive odor of an Italian restaurant (ignoring garlic for the moment).

Oregano comes into English from the Spanish *orégano*; may one cite the original word as evidence for how the word should be accented in English?

How much do we eat? In the U.S.A., it has become our leading herb; in 1995, we used 14 million pounds. One pound of oregano is enough to season pizza for a family of four every day for one year; in total, annual U.S.A. consumption of oregano equals six billion slices of pizza. These data are from the contribution of Gilbert Olivier, pp. 142–146. He is an executive in an American spice-importing company. His speech is reproduced verbatim, and there is a most refreshing immediacy to the language, almost as if he were standing before the reader.

And there's another, by Domenico Chiapparo, pp. 151–152. He tells us about trying to commercialize growing of *Oreganum heracleoticum* on his farm in Sicily. It is not a very big farm, and he is not a wealthy man. He tells what works, and what doesn't, in a charmingly spare, blunt English that I have to believe was equally appealing in the original Italian.

The conservation biologist will find much of interest here. There are a number of species of *Oreganum* that are steno-endemics; they may turn out to be critical in breeding a better commercial product, but first we have to make sure we don't destroy their natural populations. Several chapters make it clear, there is no unanimity on what the species are, despite the existence of a fairly modern monograph; this confusion will tend to stifle conservation efforts.

I find only Padulosi credited with the editing. Suffice it to say, the English is absolutely colloquial, though the authors are variously Italian, German, Turkish, Greek, Albanian, Israeli, Slovenian, and Austrian. As is typical of this series, there are very abundant and up-to-date references, and even a number of color photographs, but there is no index. Oregano has some insecticidal uses, and right at this point I would like to go back and double-check, but I can't find the references now.

Neil A. Harriman
Biology Department
University of Wisconsin–Oshkosh
Oshkosh, WI 54901, U.S.A.

Catálogo de Plantas Utiles de la Sierra Norte de Puebla, México, Cuadernos 27. Miguel Ángel Martínez Alfaro, Virginia Evangelista Oliva, Myrna Mendoza Cruz, Gustavo Morales García, Guadalupe Toledo Olazcoaga, and Alfredo Wong León. 1995. 04510. Universidad Nacional Autónoma de México, Instituto de Biología, UNAM, Apartado Postal 70–233. México, D.F., México. 303 pp. (paperback). $40.00, Mex. Pesos 210.00. ISBN 968-36-4351-6.

Most ethnobotanical books focus on one ethnic group, a specific use of plants within a society, or plant use from a cultural perspective. *Catálogo de Plantas Útiles de la Sierra Norte de Puebla* is different. This book provides fewer anthropological descriptions and more botany. Indeed, I have never before seen a guide book that is as comprehensive as this on basic ethnobotany. The *Catálogo* has an incredible amount of information on plant use among five distinct cultures, making it a unique and essential reference for cross cultural comparison in the field of ethnobotany. This is more than a checklist of useful plants; it is a practical ethnobotanical guide of the Northern Sierra of Puebla.

The *Catálogo* is compiled from data taken in various locations throughout the Northern Sierra over the last 16 years. Most information comes from Mexican student theses and social work projects that were completed in Nahua, Mestizo, Otomí, Tepehua, and Totonac communities. The book reflects the diverse cultures and flora of the Mexican states of Puebla, Veracruz and Hidalgo. Species mentioned are found in communities ranging from subtropical to pine and oak forests. Ethnobotanical information comes from 111 families, 603 species, and 398 genera of plants and fungi. Information is written in Spanish, and is organized alphabetically by family, then genera and species. The book includes helpful data such as the proportion of native and exotic species, the proportion of species by use category, the most commonly employed species, and families with the largest number of useful species. Each plant has its local common name in Spanish and its respective indigenous name. Uses, parts used, growth sites in the Sierra, vegetation type (based on Rzedowski's classification), classification (i.e., hot and cold), abundance, life form, origin, value (either for household use or market), preparation by locality, level of domestication, nutritional value, chemical composition, and finally observations also are given for each plant.

It would have been useful to have included synonyms of plant species, since some of the names in the book are not up to date. There also are a few misidentified species. However, despite these errors, I like the presentation of the book; it is practical not only for Spanish reading investigators, but to local bilingual people who can easily find plants through the appendices, and compare uses and preparations. The glossary also is a good addition since there are several illnesses not typically recognized outside of the indigenous groups represented.

The *Catálogo* is a useful book for anyone interested in Mexican ethnobotany. Although it focuses on plant use in the Northern Sierra of Puebla, many of the plants in this book are found throughout the Mexican republic, especially in the markets. I used this book throughout my research in the Sierra and found it an essential guide that is easy to follow and more than complete. I would advise anyone planning ethnobotanical research in the area to get this book, for they will find it invaluable.

Cristina A. Ugarte
Naturalist Services,
3701 S.W. 70th Ave.
Miami, FL 33170

African Ethnobotany: Poisons and Drugs. 1996. Hans Dieter Neuwinger. Chapman & Hall, Postfach 10 02 63, D-69442 Weinheim, Germany. 1996. 941 pp. (hardcover). DM 248, sFr 236, ÖS 1934. ISBN 3-8261-0077-8.

This handbook represents an important and comprehensive review of 269 plants (170 genera and 63 families) used in Africa as ingredients for arrow-hunting poisons. Among these are highly toxic plants, acting as base poisons. Many of these taxa are also renowned traditional medicinal species. All entries are described in individual monographs and listed alphabetically by family and genus.

Every monograph starts with a small map of Africa shaded in those countries where the plant is used for a hunting poison. The text is divided into: botany, vernacular names, hunting poison, traditional medicine, chemistry, pharmacology/toxicology, literature. The item "botany" comprises a short characterisation of the habit and of the different plant organs. Furthermore, the distribution and the habitat is mentioned. For almost all plants a black and white illustration is added, some-times also a distribution map. Color photographs of 59 species are included in seven plates. The list of the "vernacular names," often of surprising extent, is divided into countries and subdivided into areas where the individual names are used. "Hunting poison" comprises the description of the methods of preparation in detail along with the base poisons, e.g., the cardiac poisons of *Acokanthera, Parquetina,* and *Strophanthus*. For additional plants the preparation is only briefly mentioned. When appropriate, "ordeal poisons" is integrated. "Traditional medicine" gives short comments on the therapeutic indications, according to their use in different countries and/or tribes. "Chemistry" reproduces the secondary metabolites of the plant in a detail, including all knowledge on the phytochemistry of the specific plant. Even quantitative details are mentioned. Chemical structures of most of the constituents are presented. However, chemical structure is separated from the item "Chemistry." This is not a problem if there is only one species. If there are genera with more than one species, however, the structures are located at the end of the last species of this genus.

This placement causes two problems: (1) If the constituents of the different species represent chemotaxonomic features of such a genus, e.g., cardiac glycosides of *Strophanthus* (13 species), the structures of these compounds are given altogether (pp. 192, 193), but previously mentioned on page 157 (first *Strophanthus* species), or elsewhere between these pages (further *Strophanthus* species). Cross references at the specific item "chemistry" of a species (e.g., "for structures see page X"), would have been helpful. (2) If the constituents are only features of one species which is not the last entry of this genus, e.g., the indolizidine alkaloids of *Ipomoea,* it becomes confusing. Therefore it would be advantageous to add an underline to the chemical structures with specific coordination to the species involved.—Occasionally small trivial misprints have crept in, e.g., in the chemical structures of convallatoxin (p. 738) or ipalbidinium (p. 766).

Furthermore, there are special chapters on "Larva Poison of the Bushmen" (among other things: relationship between arrow poison larvae and their host plants) and on "Fish Poisoning Plants." Finally, a useful list on "Literature about Traditional African Medicine" is given as is a voluminous, detailed "Plant Index" and "Subject Index."

The negligible shortcomings should not overshadow the significance of this well-researched and well-documented handbook. This is a scientific treasure-house of ethnobotanical information based on profound knowledge. Neuwinger acquired the information during almost three decades of African field investigations and combined it with information from the literature. This manual with its extraordinarily detailed monographs is highly recommended as an excellent reference book to ethnobotanists and to ethnopharmacologists, pharmacists, phytochemists, medical doctors, and veterinarians interested in African medicinal and/or poisonous plants.

ECKART EICH
FREIE UNIVERSITÄT BERLIN
INSTITUT FÜR PHARMAZIE II–
PHARMAZEUTISCHE BIOLOGIE–
KÖNIGIN-LUISE-STR. 2+4
D-14195 BERLIN, GERMANY

Principles of Population Genetics, Third Edition. Daniel L. Hartl and Andrew G. Clark. 1997. Sinauer Associates, Inc., P.O. Box 407, Sunderland, MA 01375-0407. xiv + 542 pp. (hardcover). $56.95. ISBN 0-87893-306-9.

It was with no small degree of anticipation that I began my examination of the third edition of this book. The previous edition of this text (Hartl, D.L. and A.G. Clark. 1989. *Principles of Population Genetics 2nd Edition.* Sinauer Assoc., Inc. Sunderland, Mass) remains a valued component of my course library.

The first thing that was apparent was the more streamlined (542 pp. vs. 682 pp.) format of the 1997 edition. Close examination reveals a well organized, lucid, and informative treatment of the concepts of population genetics required to meet the goals specified by the authors for their text. The new edition has undergone a positive and effective re-organization of those topics. For example the *Ecological Genetics and Speciation* chapter of the earlier edition are now incorporated, in part, within the *Darwinian Selection* section of the new release. The current text begins with the requisite genetic and statistical facts, expands those concepts into genetic variability and populations before moving into advanced topics (e.g., genetic drift and selection). There is an increased botanically specific treatment of material including discussion of compensatory substitutions, fixation index, duplications, and an expanded assortative mating treatment.

The various changes have achieved a better text for a one semester course provided at the senior undergraduate level. The reorganization of material should enhance the instructional use of this text as a course book, providing better continuity among chapters than the previous version. The streamlining has come at the cost of expanded treatment of some marker systems (e.g., microsatellites, but certainly others) whose application and importance to the field have advanced considerably since the 2nd edition. There are also about thirty fewer student chapter problems, and a generally less current literature citation section than the previous edition. However, with their *Principles of Population Genetics 3rd Edition,* Hartl and Clark have successfully corrected organizational shortcomings of the previous edition and offer a good choice for advanced undergraduate courses.

MICHAEL R.J. FORSTNER
DIVISION OF SCIENCE–DEPARTMENT OF BIOLOGY
FLORIDA ATLANTIC UNIVERSITY
DAVIE, FL 33314-7714

The Biology of Rarity. Causes and Consequences of Rare-Common Differences. William E. Kunin and Kevin J. Gaston (eds.). 1997. Chapman & Hall, 2–6 Boundary Row, London, SE1 8HN, United Kingdom; 115 Fifth Avenue, New York, NY 10003 USA. xiv + 280 pp. (hardcover). $83.00. ISBN 0-412-63380-9.

Reviewing a book on the biology of rarity may seem out of place for the journal *Economic Botany.* However, those familiar with the subject, or who read this book, will discover that humans are ***the*** primary cause of rare species. In fragmentation and outright elimination of habitats the "wise human" (*Homo sapiens*) has proven to be the antithesis of its name.

This book is divided into three parts. Part one discusses "Rarity and Rare-Common Differences." The four chapters by three authors review the literature and issues involved with rare and common species. Part two covers the "Mechanisms Creating Rare-Common Differences" in eight chapters by nine more contributors. Finally, part three makes suggestions about "Future Directions." Two new contributors include ideas here, and the editors provide the concluding comments. Thus, a total of 14 contributors produced this slim and expensive volume.

Much of the text (pp. 63–236) deals with theoretical and synthetic approaches where authors seek common causal factors running between rare species. Few interconnecting strands are found, and each chapter elaborates (sometimes complains) about the overwhelming number of variables discovered. Some of these variables result from different human definitions of the same words (e.g., rarity), and others are inherent in the individual taxa studied.

The impression given by the book is that the authors forgot that they were studying individual species with different evolutionary histories, and consequently different solutions to similar problems. Instead, most chapters seem to suggest that their authors hoped to find a few common "rules" similar to those in Math, Physics or Chemistry. Almost every chapter has a section where the authors momentarily remember that they are dealing with living organisms, but more often they retreat back into the cocoon of theory enveloping the "hard sciences." The differences between the organisms are presented more often as an irritation and stumbling block to "real" science than fascinating avenues for exploration.

The only notable relief from this myopia is by Peter Cotgreave and Mark Pagel (Chapter 13). Their statement conveys much of the book's disappointment in a purposefully facetious sentence about the habitat-restricted, plant-eating Australian ground parrot (*Pezoporus wallicus*). This bird was rare but not endangered in 1986; now it is thought by the IUCN to be on the brink of extinction. The two authors conclude (p. 260, last sentence), "If only . . . [the parrot] had known about the comparative method, it could have recognized the value of eating meat or nesting in trees, and saved itself."

These twinklings of cognizance, off and on, are understandable because all the contributors deal with rare species that they fear will become extinct within their lifetimes. The authors are genuinely interested in diverting the specter of extinction. Failure to find simple, broad-reaching themes and easily-applied rules to save species surely was frustrating. Nonetheless, this is a fascinating book about how humans and myriad other factors intermingle to affect the existence of organisms on planet Earth.

DANIEL F. AUSTIN
DEPARTMENT OF BIOLOGICAL SCIENCES
FLORIDA ATLANTIC UNIVERSITY
BOCA RATON, FL 33431

Promoting the Conservation and Use of Underutilized and Neglected Crops. 12. Tef. *Eragrostis tef* (Zucc.) Trotter. Seyfu Ketema. 1997. International Plant Genetic Resources Institute (IPGRI), Via delle Sette Chiese 142, 00145 Rome, Italy. 50 pp. (paperback). Gratis. ISBN 92-9043-304-3.

Tef as a human food is raised almost entirely in Ethiopia; it has apparently been planted elsewhere, but mostly for animal feed. Cultivation of tef in Ethiopia dates to well before the beginning of the Christian era, but there is considerable uncertainty as to its antiquity beyond that broad statement.

The grains are ground into flour which is mainly used for making a popular pancake-like local bread called *enjera.* Where I come from, pancakes are made with baking powder, so they rise as they cook; I am not sure if *enjera* is leavened or unleavened, but I suspect the latter.[2] And of course the grain can be fermented to make alcoholic drinks.

The plant is annual, the flowers chasmogamous, but the plant is almost obligately self-pollinated, with outcrossing estimated at 1% or less; this at least guarantees seed set, so necessary in an annual. Clearly, there are problems for the plant breeder, however, in dealing with such tiny florets, and there is an elegant discussion of exactly how to proceed to get genes from one strain to another. Readers of this review are unlikely to be crossing tef, it is true, but there are some clever tricks in here that might be adaptable to a variety of other plants I think it merits a close look.

The book includes the obligatory analyses of chemical composition of the grains, including amino acids. Tef is low in lysine, like so many grains, but better than maize or sorghum. It is common in some parts of Ethiopia to add fenugreek (*Trigonella foenum-graecum* L., Apiaceae) to the flour, and this has the effect of making the bread softer and shinier in appearance; in addition, it supplements the lysine. Flora of Ethiopia 3: 246. 1989, says that fenugreek is widely cultivated in the country as a spice or condiment in baking; the *enjera* not only tastes better, it's better for you.[3]

Ethiopian farmers get a reasonable crop on a rainfall regime of 300 mm per year; that works out to roughly 12 inches of rain a year, a very arid situation indeed, especially given the rapidity with which the moisture must evaporate in that blazing African sun. Nonetheless, the yields (kilogram per hectare) are better than for maize, barley, sorghum, and wheat in Ethiopia.

Tef has few serious pests, as one would expect; if you are an annual native grass, you have to have evolved to be in balance with the fungi and insects that are waiting to set upon you.

Like all the volumes in this series, the book is provided with stiff, shiny covers and sewn signatures on good-quality paper. Ms. Linda Sears is mentioned and thanked on p. 5 as the editor who brought the entire work into its present fine form. She and the author are to be congratulated.

NEIL A. HARRIMAN
BIOLOGY DEPARTMENT
UNIVERSITY OF WISCONSIN–OSHKOSH
OSHKOSH, WI 54901, USA

[2] Editors Note: My colleagues & friends Sebsebe Demissew and Nigist Asfaw of Addis Ababa, Ethiopia, inform me that, while baking powder is not used in Ethiopia, there is a leavening agent. That material is a "starter" yeast called ersho remaining from a previous portion of the fermented batter.

[3] Editors Note: Sebsebe Demissew and Nigist Asfaw say that this practice is declining in some parts of Ethiopia, especially around Addis Ababa.

Species. The Units of Biodiversity. M. F. Claridge, H. A. Dawah, and M. R. Wilson. (eds.). 1997. The Systematics Association Special Volume Series 54. London SE1 8HN, United Kingdom, Chapman & Hall, 2–6 Boundary Row. xvi + 439 pp. (hardcover). U.S. $89.95, £49.00. ISBN 0-412-63120-2.

Volumes have been written on species concepts. According to this book, more pages have been devoted to the topic than any other in the evolutionary sciences. This new book follows in the same vein, with a difference—the authors of this work cover organisms from viruses to mammals and plants. Most authors present the strengths and weaknesses of the concepts being used—and all 22 species definitions have failings. Indeed, Cracraft (Chap. 16) suggests that looking for the ideal species concept may be comparable to the search for the Holy Grail.

Throughout the book several authors make reference to biologist's desire to emulate the "hard sciences" to discover universally applicable "rules" of Nature. I cannot help remembering Bernd Heinrich's comment that, "In physics, if it works, it is generally considered correct. But evolution has jury-rigged organisms to work in all sorts of ways . . ." (from *Ravens in Winter*, 1989. Vintage Books, New York). Although Heinrich was not talking about species at the time, this observation applies.

Species consists of 19 chapters. The discussion is ably introduced by the editors (Chap. 1, "Practical approaches to the species concepts of living organisms") and all views skillfully critiqued by David Hull (Chap. 18, "The ideal species concept—and why we can't get it"). The last chapter (Chap. 19) by R. L. Mayden suggests that there should be "A hierarchy of species concepts: the denouement in the saga of the species problem." Between these are discussions of viral species (Chap. 2 by M. H. V. Regenmortel), cultivable bacteria (Chap. 3 by M. Goodfellow, G. P. Manfio and J. Chun), uncultured prokaryotic diversity (Chap. 4 by T. Martin Embley and E. Stackebrandt), eukaryotic algae (Chap. 5 by D. M. John and C. A. Maggs), lichens (Chap. 6 by O. W. Purvis), fungi (Chap. 7 by C. M. Brasier), plants (Chap. 8 by R. J. Gornall), cultivated plants (Chap. 9 by J. G. Hawkes), marine invertebrates (Chap. 10 by N. Knowlton and L. A. Weigt), nematodes (Chap. 11 by D. J. Hunt), herbivores and parasitoids (Chap. 12 by the editors), blood-sucking vectors of human disease (Chap. 13 by R. Lane), parthenogenetic insects (Chap. 14 by R. G. Foottit), terrestrial non-insect invertebrates (Chap. 15 by A. Minelli and D. Foddai), and mammals (Chap. 17 by G. B. Corbet).

All biologists will find useful practical and theoretical information in each chapter. Some of the ideas seem to be based more on politics than science (cf. Chap 16), although their authors would doubtless disagree. If you do not read any other part of the book, read Chapter 19. There Mayden uses the Evolutionary Species Concept to put everything in a theoretical context comparable to the "hard sciences." This is a readable compilation with current references. *Species. The Units of Biodiversity* should be read by all who deal with species, in either evolutionary, systematic, ecological, ethnobiological, or other contexts.

DANIEL F. AUSTIN
DEPARTMENT OF BIOLOGICAL SCIENCES
FLORIDA ATLANTIC UNIVERSITY
BOCA RATON, FL 33431

Non-Wood Forest Products: Tropical Palms. Dennis V. Johnson. 1997. Food and Agriculture Organization of the United Nations, Bangkok, Thailand. RAP Publication 1997/10: Non-Wood Forest Products. viii + 166 pp. (softcover).

The world's palms stand third (behind the grasses and legumes) in "utility of the products derived from them (p. i)." All members of the family *Arecaceae* (alternately *Palmae*), are distributed across a wide range of ecological zones, yet are most prolific in the lower strata of the humid, tropical forest canopy. Johnson has compiled an extremely handy, well-organized text devoted to the conservation and economic development of this extraordinary plant family. In writing the book he used data acquired from the World Conservation Monitoring Centre (WCMC), the palm specialist group of the World Conservation Union/Species Survival Commission, the voluminous bibliographic sources on the human use of palms, and his own extensive research.

Palms have a unique association with human beings. For millennia they have clothed, sheltered, nurtured, intoxicated, healed, adorned, and entertained the human species. Yet, as they are sometimes equated with weedy, secondary-growth trees, some are under considerable threat from the activities of humans themselves. Johnson's text was written to help ameliorate this problem and has succeeded in becoming an invaluable resource for the conservationist, protected-areas manager, or tropical forester. Ethnobotanists may find some of the chapters oversimplified, but they will be pleased as well, especially with the lengthy reference section.

The book's nine chapters are divided into three thematic sections: 1) a botanical introduction, a discussion of the role of palms in human culture, and a list of current palm products; 2) a geographical discussion of the non-timber uses of palms in Asia, the Pacific, Latin America, and Africa; and 3) a conclusion devoted to palms with exceptional development potential. The chapters that precede the geographical discussion of palms are the weakest chapters of what is otherwise a very strong book. Although brevity was an obvious consideration given the books intended audience, the reader may become frustrated with the lack of depth in the early subheadings "Case Studies: Indigenous Groups and Their Use of Palms" and "Palm Domestication."

The strength of the text is certainly in the section devoted to exploring the current non-timber uses of palm species over the four geographic areas. Each geographical subsection consists of an accessible table documenting threatened and non-threatened species, the common names associated with those species, and the distribution of the plants within the region. These tables are accompanied by brief discussions highlighting the conservation status of individual species and their potential as a commercial resource. Color photographs accent the human use of palms for each geographical subsection.

Johnson then briefly discusses 21 palm species that—either through an increased effort in domestication or management—have the greatest promise in economic development plans. The various species of "rattan" (*Calamus* spp.), the "babaçú" of South American tropical rain forests (*Orbignya phalerata*), the semi-desert African "doom" (*Hyphaene* spp.), and the SE Asian "salak" (*Salacca zalacca*) all make the list. Twenty five tables documenting the nutritional composition of the palms selected for development potential serve as a conclusion.

Johnson has served the conservation and sustainable development readership well with the publication of *Tropical Palms.* Those more familiar with the botany and anthropology of palms will certainly welcome the text as an excellent compendium.

DARRON ASHER COLLINS
DEPARTMENT OF ANTHROPOLOGY
TULANE UNIVERSITY
NEW ORLEANS, LA

Weed Ecology. Implications for Management. 2nd edition. Steven Radosevich, Jodie Holt, and Claudio Ghersa. 1997. John Wiley & Sons, Inc., 605 Third Avenue, New York, NY 10158–0012. xvi + 589 pp. (hardcover). $89.95 in USA; $125.95 in Canada. ISBN 0-471-11606-8.

According to these authors, 250 species of plants are considered weeds throughout the world. They calculate that this is 0.1% of all plants. However, they point out that 75% of the world food production comes from 12 crops. Thus, humans owe their lives to a minuscule number of plants—12 of 250,000 or 0.0048% of those known. Those cultigens are subject to many competing species. What makes these data more frightening is that, ". . . the abundance of both new and existing weed species increased during the last half century" (p. 531). This increase comes at an expense of millions of dollars to throw at weeds every technology known to humans—from the ancient hoe, to chemicals, biotechnology, and biological control. Do you get the idea that we are doing something wrong? I do!

Parts I and II deal with basics. Chapter 1 is on "Weeds and Weed Science." Here fundamental definitions are given of "weeds" as they occur in agricultural land, forests, and rangelands. Also, a brief history of weed science is given. Next are the "Principes of Weed Ecology, the "Genetics and Evolution of Weeds," "Weed Demography and Population Dynamics," and "Associations of Weeds and Crops." Then there is a shift to "Physiological Aspects of Competition" and "Other Types of Interference" such as allelopathy, parasitism, commensialism, and mutualism. Part III deals with the "Technology of Weeds Sciences." This includes "Methods and Tools of Weed Management," herbicides, and the action and fate of herbicides. Finally, Chapter 11 discusses "Weed Control in a Social Context." The Epilogue "What Is a Weed?" should, in my opinion, have been the introduction to the book. Last are the references, two appendices, and an index.

The authors say this volume has been updated with ideas, data, and publications over the past four decades. Indeed, many of the numerous references cited (pp. 535–568) are dated 1996 when this book came off the presses. Presumably, this book is like the first edition in that it is aimed at a mixed audience including undergraduates and graduates. Some of the introductory chapters are written for people illiterate in the basics of biology and ecology.

Superficially, the introductory chapters appear well-founded. Indeed, there is nothing wrong in what is said, but more careful reading shows that the ecological concepts are based mostly on the research of four people: Herbert G. Baker, Jack Harlan, John Harper, and Edward Salisbury. The authors seem completely innocent of the modern views of "weeds" existing apart from land devoted to AgroIndustry. There is no mention, for example, of the recent books on alien plant invasions (e.g., McKnight 1993; Pysek et al. 1995; U.S. Congress 1993). Impacts of exotic "weeds" on wild areas are not considered.

This book is replete with information at several levels regarding weed ecology. For those interested in any aspect of weeds and humans, this is a good place to begin.

Literature Cited. *(1)* McKnight, B. N. (ed.). 1993. Biological Pollution: The Control and Impact of Invasive Exotic Species. Indianapolis, Indiana Academy of Science. *(2)* Pysek, P., K. Karel Prach, Marcel Rejmánek, and Max Wade (eds.). 1995. Plant Invasions. General Aspects and Special Problems. Amsterdam, The Netherlands, SPB Academic Publishing. *(3)* U.S. Congress. 1993. Harmful Non-Indigenous Species in the United States. Washington, DC, Government Printing Office, OTA-F-565.

Daniel F. Austin
Department of Biological Sciences
Florida Atlantic University
Boca Raton, FL 33431

Oryza: From Molecule to Plant. Sasaki, Takuji and Graham Moore (eds.). 1997. Klumwer Academic Publishers, P.O. Box 989, 3300 AZ, Dordrecht, The Netherlands. Reprinted from *Plant Molecular Biology* 35 (1,2). 1997. 254 pp. (hardcover). U.S. $148, NLG 250, £90. ISBN 0-7923-4455-3.

"The population of rice consumers is increasing at the rate of 1.8% annually. The present annual rice production of 560 million tons must be increased to 850 million tons by 2025 (p. 33)." Why this was calculated to 2025 is not explained. A further bit of arithmetic says that, if this rate of growth continues, then that population will double in 39 years; therefore, 39 years from 1997, in the year 2036, we will need 1120 million tons. It is further pointed out in the preface that we **will need as much rice and wheat in the next 50 years as has been produced since the beginning of agriculture**—there is little time to waste. How all these numbers mesh is not clear. Suffice it to say, the doomsday prophets may turn out to be right after all, just off in their predictions by a few decades, and a tragedy of unimaginable proportions may yet face our species.

The editors (and surely, all the authors) hold the view that meeting such production goals can only be possible if biotechnology is allied with more traditional, classical breeding programs. The molecular approaches detailed here will permit the introduction of useful genes of very different races and species into cultivated rice. Indeed, the translation of laboratory work into breeding programs has already begun.

As one would gather from the fact that this is a reprint of a technical journal, much of this book is quite impenetrable to the general reader. Its audience is the gene cutters themselves. I would like to know what synteny, microsynteny, and gene golfing mean (ask your local genetics professor, as I did), but this is not a primer for the neophyte. Nonetheless, a few of the articles are accessible, such as "Origin, dispersal, cultivation, and variation of rice," (pp. 25–34), a very useful summary, and "Conservation of rice genetic resources . . ." (pp. 61–67), followed by "Rice genetic resources . . . in Japan," (pp. 69–77).

A most important point is also made. The molecular techniques applied here to rice will work equally well on grasses in general, which should make this book appealing to many different kinds of plant breeders.

Wheat is given more than passing mention, for example, in a paper on comparative genetics of grasses.

The entire work is lavishly produced on high-quality paper with full-color illustrations here and there. It ought to be, given the price.

NEIL A. HARRIMAN
BIOLOGY DEPARTMENT
UNIVERSITY OF WISCONSIN–OSHKOSH
OSHKOSH, WI 54901 USA

Biotechnology and Plant Genetic Resources, Conservation and Use. J. A. Callow, B. V. Ford-Lloyd and H. J. Newbury (eds.). 1997. CAB International; distributed by Oxford University Press, 198 Madison Avenue, New York, NY 10016-4314. X + 308 pp. (hardcover). $90.00 ISBN 0-85199-142-4.

The editors say the inspiration for this volume was the revolutions in biotechnology for developing novel molecular and cellular technologies, and computer information and networking. The contents are broad: use of molecular marker techniques for description of plant genetic variation; genetic diversity—population structure and conservation; genomic relationships, conserved synteny and wide-hybrids; molecular markers and the management of genetic resources in seed genebanks: a case study of rice; in vitro conservation methods; conservation of DNA: DNA banking; genetic resources and plant breeding; gene identification, isolation and transfer; importance of biotechnology for germplasm health and quarantine; biodiversity for bioindustries; and internet resources for the biologist.

The editors also state that, "Where traits are controlled by one or a few genes, direct isolation of genes and their incorporation into elite cultivars by increasingly efficient transgenic technologies (Chapter 9) provide an additional approach to the more traditional methods of plant breeding. This prospect also challenges one of the classical concepts in plant genetic resources work, namely the 'gene pool' concept of Harlan and de Wet (1971), which separated cultivated species and their wild relatives into different categories based upon their 'crossability.' The value of this concept in guiding the work of the plant breeder must now be re-evaluated in the light of molecular genetic technologies which have not only cast a completely new light on the relatedness plant genomes (Chapter 4), but which have effectively widened the potential gene pool of a species to include all other life forms." Here, indeed, is another revolutionary outcome.

Chapter 2, on the use of molecular markers for description of plant genetic variation, is a comprehensive survey of this rapidly-growing realm, as DNA sequence information is the foundation for developing and applying markers to biological variation. The authors claim that ". . . relationship studies will continue to benefit greatly from information and technologies generated by the Human Genome Project." Also that, "Notable achievements—now taken for granted by many—include new types of genetic markers (particularly those based on PCR [polymerase chain reaction]), more efficient cloning vectors, the introduction of STS [not defined] markers, and improved technology and automation for DNA sequencing." This chapter is crammed with useful information, but the excessive use of acronyms sent me searching for explanations.

A major issue in genetic resources for some time has been the size of germplasm collections for their effective management and use. Sir Otto Frankel suggested in 1984 a core collection—a subgroup of accessions that incorporates, with minimum redundancy, the genetic diversity of a crop species and its relatives. Chapter 5, on molecular markers and the management of genetic resources in seed genebanks, informs us that PCR-based molecular markers are increasingly used to assess genetic diversity in germplasm collections. "The advent of molecular characterization and evaluation of germplasm opens another chapter in genetic conservation, and one which will fundamentally change our perspectives on the nature, structure and value of crop gene pools."

This volume should be in every library concerned with plant genetic resources. Although the articles vary in presentation and complexity, the information is essential to practitioners and educators who wish to integrate genetic theory into conservation strategies.

DOROTHEA BEDIGIAN
YELLOW SPRINGS, OH

Tiwanaku and its Hinterland. Archaeology and Paleoecology of an Andean Civilization. 1. Agroecology. Alan L. Kolata (ed.). 1996. Smithsonian Institution Press, 470 l'Enfant Plaza, Suite 7100, Washington, DC. 20560. xix + 323 pp. (hardcover). $95. ISBN 1-56098-600-X.

This is the first of a two-volume set reporting on a six year investigation into Tiwanaku, a civilization that developed high in the Bolivian Andes. This volume aims to demonstrate that raised-field agricultural techniques enabled the people of Tiwanaku to produce enough food to sustain an urban population in a climatically-marginal zone, in the first millennium AD.

In chapters two through eight, agroecological parameters of the raised-field system are established through detailed studies of the physical environment in the river valleys which drain into the southeastern shore of Lake Titicaca. The applied aspect of the project, rehabilitating ancient raised-field systems, is outlined in chapters nine through eleven. Problems encountered during the rehabilitation process underscore the critical role of social organization in determining the relative success of raised-field systems.

Kolata sandwiches the chapters on environmental and experimental data with introductory and concluding chapters addressing theoretical issues surrounding the emergence, maintenance, and collapse of states. He acknowledges that Wittfogel's (1957) claim that centralized authority is necessary for large-scale irrigation is not always true, but he cautions that the belief that corporate kin groups can maintain vast tracts of raised-fields on their own may be equally flawed. By way of compromise, he suggests that although the first Tiwanaku raised-fields were probably established by small groups, the canal systems built from AD 400 to 1000 could only have been designed and maintained by a centralized elite.

To explain the fall of Tiwanaku, interpretations of glacier and lake cores are used to argue that aridification commenced about AD 1000, lowering ground water levels to the point where raised-field technology no longer worked. Without surplus food production in the rural zone, the urban centers collapsed. Curiously, construction and use of one raised-field is dated to AD 1230, and another to AD 1300 (Table 8.1). Moreover, two of four "problematic" radiocarbon dates range from AD 1310 to 1480. Therefore, it is perplexing that Kolata dismisses Graffam's (1990) thesis that raised field construction continued after the collapse of the Tiwanaku urban centers.

Their disagreement bears on planning for future management strategies of raised-field systems. If Kolata is right, the Aymara corporate kin groups (*ayllus*) will not be able to maintain the rehabilitated raised-fields in the absence of an over-arching bureaucracy, but if Graffam is right, they should be able to do so on their own. A continuation of Alice Kehoe's observations of Aymara social dynamics associated with rehabilitating the raised-fields may help resolve this issue.

The book's interdisciplinary nature ensures that it will be of interest to a wide range of professionals, yet its price may discourage non-archaeologists from buying it. However, for those interested in cultural ecology, Andean civilizations, and comparative studies of pre-Industrial states, this volume has much to offer.

Literature Cited. *(1)* Graffam, G. 1990. Raised Fields without Bureaucracy: An Archaeological Examination of Intensive Wetland Cultivation in the Pampa Koani Zone, Lake Titicaca, Bolivia. Unpublished Ph.D. dissertation, University of Toronto. *(2)* Wittfogel, K. 1957. Oriental Despotism: A Comparative Study in Total Power. Yale University Press, New Haven.

KEVIN LEONARD
MOUNT ALLISON UNIVERSITY
144 MAIN STREET
SACKVILLE, NB E4L 1A7
CANADA

Soil, Water, Biology and Belief in Prehistoric and Traditional Southwestern Agriculture. H. Wolcott Toll, (ed.). 1995. New Mexico Archaeological Council, P.O. Box 1023, Albuquerque, NM 87103. xvi + 373 pp. (paperback). $25.00. ISBN 0-964931-1-9.

Most of these papers were presented at an agriculture symposium held by the in Santa Fe New Mexico Archaeological Council in October of 1992. The participants represented interdisciplinary interests: the core group was Southwestern archaeologists with active interests in the study of field systems and irrigation, and archaeobotanists who frequently work with archaeologists. The "heartening expansion of scope" came with the presence of historians, architects, agronomists, biologists, geographers and engineers, and Native American and traditional farmers.

The volume is divided into four sections: *Field Systems, Soil, Water,* and *Biology and Belief.* The section on *Field Systems* includes "A comparative study of prehistoric farming strategies"; "Anasazi field systems in Taos District"; "Diverse agricultural technologies to harvest and conserve water in Lower Rio Chama Valley, New Mexico"; "14th century dry farming features in northern Rio Grande Valley"; "Estimating agricultural carrying capacity in the Zuni region"; and "Mixed agricultural technologies in southern Arizona."

The first article in the *Soil* section is "Searching soil for clues about southwest prehistoric agriculture," by Jonathan A. Sandor. This is noteworthy since the germinal idea for the symposium was planted by David Love, a geomorphologist, who suggested a meeting on agriculture including Sander would be exciting. His paper focuses mainly on those agricultural systems that relied on rainfall and runoff, rather than formal irrigation. Two areas of emphasis are soil and landscape settings for prehistoric agriculture, and effects of agricultural land use on soil. Strategies for runoff agriculture are inferred from topographic and soil placement of fields. The other titles are "Biological nitrogen fixation and soil fertility: implications for Anasazi Agriculture"; Water availability for plant growth in pre-Columbian terrace soils, Chihuahua, Mexico"; and "An interpretive framework for radiocarbon dates from soil organic matter from prehistoric water control features."

The section on *Water* includes "The case for prehistoric irrigation in northern southwest"; "Climatic overview of the Rio Grande del Rancho Valley: background for irrigation"; Prehistoric irrigation at Pot Creek Pueblo"; "Prehistoric irrigation in the Taos Valley"; "Mogollon/Western Pueblo soil and water control systems of the reserve phase"; and "San Saba-Menard Irrigation system: origins, important variables for prehistoric irrigation studies."

Biology and Belief contains "Traditional agriculture

on the Zuni Indian Reservation"; "Maize productivity and agroecology: effects of environment and agricultural practices on the biology of maize"; "Agriculture and the Mesa Verde Anasazi diet"; "Pollen evidence of prehistoric agriculture"; "The where, when and why of corn guardians"; and "Dryland gardening in southwest Colorado: past and present." Vorsila L. Bohrer's contribution on corn guardians suggests that selection pressure can be evaluated by comparing the morphology of the field crop with that of the most valued ears.

This is a unique book that belongs in college and university libraries, and in the personal libraries of all ethnobotanists with New World interests. It illustrates the power of interdisciplinary conferences and collaboration perfectly. Unfortunately, the binding of this volume is fragile and pages tear out easily.

DOROTHEA BEDIGIAN
YELLOW SPRINGS, OH

Nafanua: Saving the Samoan Rain Forest. Paul A. Cox. 1997. W. H. Freeman & Co., 41 Madison Ave., New York, NY 10010. viii + 238 pp. (hardcover). $23.95. ISBN 0-7167-3116-9.

From the perspective of a botanist, an anthropologist, a naturalist and an activist, Cox presents in his book an insightful contribution about his involvement in the study and preservation the rainforest of Western Samoa. He starts with his mother's death from cancer and his search for a drug to cure the disease. In this exciting and beautiful account, the reader is exposed to major issues relating to other cultures, conservation, research, and politics.

His description of the flora and ecology of the forest is so evocative that the reader can visualize it. His description of the plants and animals demonstrates the intrinsic complexity of this ecosystem. The book discusses not only the ecology of flying foxes, but also the difficulties of getting them listed as endangered species. Cox raises important issues involved in the work of ethnobotany. Moreover, he provides an excellent ethnoecological description of the relationship of the local Samoan indigenes with their natural environment. His ethnographic description is refined with a deep understanding of the Samoan culture due to his fluency in the Samoan language. Cox neither romanticizes the Samoans nor criticizes them, but provides a balanced and respectful description of them as individuals and as a society by portraying their problems and strengths. His book provides a rich cultural description that not many anthropologists can achieve. Even though Cox tried to maintain an apolitical scientific stance, he had no choice but to fight to help the subjects of his research (the forest, the rare species of flying fox and the indigenes) as an activist. This book shows what an ethnobotanist ought to be doing while conducting research.

Unlike the history of Easter Island, which ended with the destruction of its people and forest, the island of Savaii has a window to a more positive future thanks to the work and involvement of Cox. He describes how the local politics and protocol are complex and far more sophisticated than most westerners could imagine for 'primitive' societies. Cox's dedication and commitment freed this village from having to decide to sacrifice this forest for the education of their children by finding another solution.

This book shows how Cox got the funds from developed nations to save the forest and guarantee the education of Falealupo children. After achieving these solutions, he personally experienced the destruction of this village and its natural environment by a hurricane. Then, he became involved in rebuilding it. It is not surprising that he earned both the chiefly title of Nafanua and the Goldman Environmental Prize.

This book is a wonderful case study of the fieldwork, life and research of an ethnobotanist struggling to succeed in his work by dealing with the different ethical issues that most ethnobotanists face in the field. The complications of getting indigenous people's permission are sometimes out of the control of the researcher. Cox discusses the difficulties of minimizing our western cultural impact on the Samoan culture. The issue of who pays for the conservation of the forest is well presented. Even though its destruction affects the availability of resources to the local people, they prefer to improve their chance for their children to get a western education, since their future is more important.

I sincerely recommend this exciting and well written book to those interested in ethnobotany, the rain forest, Samoan culture, and field work.

MANUEL LIZARRALDE
DEPARTMENTS OF ANTHROPOLOGY AND BOTANY
CONNECTICUT COLLEGE
270 MOHEGAN AVENUE
NEW LONDON, CT 08320-4196

BOOKS RECEIVED[1]

A Dictionary of Genetics, Fifth Edition. Robert C. King and William D. Stansfield. 1997. Oxford University Press, 198 Madison Avenue, New York, NY 10016. x + 439 pp. (paperback). $24.95 (paperback); $45.00 (hardcover). ISBN 0-19-509442-5 (paperback); 0-19-509441-7 (hardcover). . . . This book received a rave review in the *Florida Entomologist* 80(3):413–416. 1997; online version http://www.fcla.ufl.edu/FlaEnt/fe803.htm. According to the propaganda in the book and the review, there are 6600 definitions of genetic, evolutionary and molecular biology terms.

A Manual of California Vegetation. John O. Sawyer and Todd Keeler-Wolf. 1995. California Native Plant Society, 1722 J Street, Suite 17, Sacramento, CA 95814. 471 pp. (paperback). $39.00. ISBN 0-943460-26-2. . . . This synopsis of the native plant covering of California is a classification system where the ". . . Focus is on Floristics and Rarity." As such it should facilitate conservation and management of the remaining native habitats in the state. All other states would profit from having an equally advanced compilation of their green resources. While anyone may find something to criticize about the system used (particularly the overabundant abbreviations on pages with ample blank space that would have allowed whole words to be spelled out), this book stands as a model for others concerned with the status of their native floras.

Coping with Risk in Agriculture. J. B. Hardaker, R. B. M. Huirne, and J. R. Ande. 1997. CAB International; distributed by Oxford University Press, 198 Madison Avenue, New York, NY 10016. xi + 274 pp. (paperback). $40.00. ISBN 0-85199-119-X. . . . A "how-to" guide for farmers to maximize their profits and minimize their losses.

Crop Evolution, Adaptation and Yield. Lloyd T. Evans. 1993 (1996 first paperback edition). University of Cambridge Press, 40 West 20th Street, New York, NY 10011-4211. xii + 500 (paperback). $34.95. ISBN 0-521-29558-0. . . . This ambitious book takes a different approach to crop evolution. Author Evans (CSIRO Division of Plant Industry, Canberra, Australia) views crops from the perspective of plant physiology. Crop yield is the pervasive theme in the chapters on origins, adaptation and improvement, germplasm diversity, plant physiology and agronomy. The message is that yield does not fit a reductionist scientific model. There are numerous examples illustrating the author's thesis. This lengthy book, filled with information that may be overwhelming, is valuable for sensitizing readers to the plethora of factors influencing crop production, and the importance of awareness of the complex ecological interrelationships undergirding the global food supply.—Mary Eubanks, Department of Botany, Duke University, Durham, NC 27708-0338.

Field and Laboratory Guide to Tree Pathology, Second Edition. R. O. Blanchard and T. A. Tattar. 1997. Academic Press, a division of Harcourt Brace & Co., 525 B. Street, Suite 1900. San Diego, CA 92101-4495. ix + 309 pp. (paperback). $19.95. ISBN0-395-79761-6. . . . This second edition of this book provides an illustrated rogues-gallery of tree diseases and other problems. This is followed by 28 pages of laboratory exercises on various maladies.

Forest Pharmacy. Medicinal Plants in American Forests. Stephen Foster. 1995. Forest History Society, Inc., 701 Vickers Avenue, Durham, NC 27701. 64 pp., 16 color illus. (paperback). Price not given. ISBN 0-89030-051-8. . . . Foster gives a history of the long battle between what finally became the American Medical Association (AMA) and alternate medicinal treatment in the USA. He compares the USA system with those in Canada, Europe and the World Health Organization. This booklet clearly shows the profit, and not humanitarian, basis for health treatment in the USA. A long list of plants historically used in the USA concludes the book.

Horticultural Therapy and the Older Adult Population. S. E. Wells (ed.). 1997. The Haworth Press, Inc., 10 Alice Street, Binghamton, NY 13904-1580. 210 pp. (paperback). $24.95. ISBN 0-7890-00458. . . . A book touting gardening to keep the older population active and usefully occupied.

Malesian Seed Plants. Vol. 1. Spot-characters. An Aid for Identification of Families and Genera. M. M. J. van Balgoody. 1997. Publication Department, Rijksherbarium—Hortus Botanicus, P.O. Box 9514, 2300 RA; distributed by Backhuys Publishers, P.O. Box 321, 2300 RA, Leiden, The Netherlands. 154 pp. (paperback). Dfl. 50.00. ISBN 90-71236-31-5. . . . This book formalizes the notes collected by C. G. G. J. van Steenis, R. C. Bakhuizen van den Brink Jr. and F. H. Hildebrand and the author made over the past 30+ years on plant

[1] Unsigned commentaries are by the book review editor; he alone is responsible for opinions expressed.

Economic Botany 52(4) pp. 441–443. 1998

collections from Malesia by publishing lists of genera with distinctive traits. There are lists of cushion plant, those with swollen nodes, with white or yellow sap, fragrances, indumentum, glands, and many others. Doubtless a useful book for students of the region.

Nematode Vectors of Plant Viruses. C. E. Taylor and D. J. F. Brown. 1997. Oxford University Press, 198 Madison Avenue, New York, NY 10016. xii + 286 pp. (hardcover). $80.00. ISBN 0-85199-159-9. . . . This text provides an in-depth look at plant viruses and their transmission via nematodes. The book includes nematode biology, taxonomy, identification, and worldwide distribution. Data dealing with all groups of plant viruses, known associations with nematodes, and suggestive evidence is presented. The role of nematodes in the life cycle of these viruses is included. Nematodes as a point of control of a number of these viral diseases is also discussed. Overall an useful reference text for individuals involved in control of plant viral diseases.—Michael Roner, Department of Biological Sciences, Florida Atlantic University, Boca Raton, FL 33431.

Plant Collectors in Madagascar and the Comoro Islands. A biographical and bibliographical guide to individuals and groups who have collected herbarium material of algae, bryophytes, fungi, lichens, and vascular plants in Madagascar and the Comoro Islands. Laurence J. Dorr. 1997. Royal Botanic Gardens, Kew, Richmond, Surrey, TW9 3AB, England. Price not given. ISBN 0-12-103982-X. . . . This volume goes far beyond what the title implies, as it includes information on many who had only a passing brush with these islands. For individuals listed, there are bibliographic references to their critical works on the islands, localities of their collections, and, in some but not all cases, a photograph of them. The book is one that will be consulted by those who wish to learn about botanists who have worked in the region.

Plants, Man and Life. Edgar Anderson. 1997. Missouri Botanical Garden, P.O. Box 299, St. Louis, MO 63166-0299. xiv + 251 pp. (hardcover). $16.95. ISBN 0-915279-44-4. . . . This book remains as delightful as it was when it first appeared in 1952. Although some of Anderson's theories have been eclipsed by newer views, his ideas remain as crisp and on-target as they were 45 years ago. Biologist or not, who would not be intrigued by titles such as "How to measure an avocado," "Science and the Bureaucrats" or "Budgets vs. Scholarship?"

Taylor's Guide to Ornamental Grasses. Roger Holmes and Frances Tenenbaum. 1997. Houghton Mifflin Co., 222 Berkeley Street. Boston, MA 02116-3764. ix + 309 pp. (paperback). $19.95. ISBN0-395-79761-6. . . . This is a mini-encyclopedia of the grasses available for cultivation, particularly ornamentals in temperate zones. The book contains over 165 color photographs of grasses in many different gardens. The authors apparently are innocent that many of these plants are rampant weeds in wild regions of North America and other parts of the world. At least, no mention is made of their escaping.

The Evolving Coast. Davis, R. A. J. 1997. New York NY 10010, Scientific American Library, distributed by W. H. Freeman Co., 41 Madison Avenue. 233 pp. (paperback). $19.95. ISBN 0-7167-6021-5. . . . This paperback, corrected edition of a 1994 book explains in layman's terms the processes and features of changing coastlines of the world. Richly illustrated, this is an excellent introduction to the structure and dynamics of coasts. One of the primary points the author makes is that *Homo sapiens* changes coasts more dramatically and rapidly than other factors in Nature.

The History and Geography of Human Genes. Abridged Paperback Edition. L. Luca Cavalli-Sforza, P. Menozzi, and A. Piazza. 1996. Princeton University Press, 41 William Street, Princeton, NJ 08540. xvii + 413 pp. (paperback). $35.00, £25. ISBN 0-691-02905-9. . . . This is a book that biologists of every ilk should read and consider. The team generated this study as a prelude to the "Human Genome Project" that has received so much publicity. This version and its hardcover predecessors have not garnered as much press because they are "just books," and not high-tech gene-sequencing technology. *The History and Geography* will certainly raise questions about all things having to do with the species that interests all *Homo sapiens*. Whether or not one agrees with the conclusions in the various chapters, this book MUST be required reading for biologists. It would be great if we could also get politicians, religious leaders, and others to read and consider its contents.

The Plant-Book. A Portable Dictionary of the Vascular Plants, second edition. D. J. Mabberley. 1997. Cambridge University Press, 40 West 20th Street, New York, NY 10011-4211. xvi + 858 pp. (hardcover). $49.95. ISBN 0-521-41421-0. . . . Those who are familiar with the first edition of this book will be delighted with this expanded and updated edition. Literature up to the first months of 1996 has been included, and an introductory page claims that nearly 2500 entries have been added to bring the total to over 20 000. Whatever you want to know about a plant, this book is the place to begin. There is a remarkable quantity of human-related information also included.

Vascular Plants of Texas. A Comprehensive Checklist Including Synonymy, Bibliography, and Index. 1997. Stanley D. Jones, Joseph K. Wipff, and Paul M. Montgomery. University of Texas Press, P.O. Box 7819, Austin, TX 78713.

xii + 404 pp. (hardcopy). $55.00. ISBN 0-292-74044-1. . . . This is a useful update of the *Manual of the Vascular Plants of Texas* by Correll and Johnston (Texas Research Foundation, 1970). By selecting out the names of plants that pertain only to Texas, the authors have made them more available than they might be in Kartesz's *A Synonymized Checklist of the Vascular Flora of the United States, Canada and Greenland* (Timber Press, 1994) or its online updates. Curiously, the authors seem to be unaware ofthe ongoing studies at Texas A&M University (http://csdl.tamu.edu/FLORA/), or apparently the Bioinformatics Working Group (http://www.csdl.tamu.edu/FLORA/tamuherb.htm) who have provided printed and web versions of the Hatch et al.'s Texas checklistfor some time (http://www.csdl.tamu.edu/FLORA/taes/tracy/regecoNF.html).

Progress in Botany. 56. Structural Botany, Physiology, Genetics, Taxonomy, Geobotany. Behnke, H.-Dietmar, Ulrich Lÿttge, Karl Esser, Joachim W. Kadereit, Michael Runge (eds.). 1995. Springer-Verlag New York, Mercedes 160 Imlay St., Brooklyn, NY 11231. xvi + 490 pp. (hardcover). $214.00. ISBN 3-540-58407-2. . . . As with the others in this series, this book summarizes the changes in botany, focusing on 1993–1994. The subtitle gives a summary of the topics covered except for the introductory review by Hubert Ziegler who discusses "Stable Isotope in Plant Physiology and Ecology."

VOLUME 52: INDEX TO SUBJECTS[1]

[1] Single page numbers indicate only that an entry occurs in an article, not its frequency.

Economic Botany 52(4) pp. 444–458. 1998

VOLUME 52: INDEX TO AUTHORS AND TITLES OF PAPERS

VOLUME 52: INDEX TO MANUSCRIPT REVIEWERS 1998

Joanne Abbot
Kathryn S. Anderson
Renée Ankarfjärd
Mohamed Bakarr
Ganasan Balachander
William L. Balée
Jim Bauml
K. S. Bawa
Mauricio Bellon
John Beutler
Eric Boa
David Bosher
Javier Caballero
Bruce Campbell
Paolo Casoria
Jeanette Clark
A. B. Cunningham
Ardeshir B. Damania
Deena Decker-Walter
Timothy Dickinson
Mark Dimmitt
Memory Elvin-Lewis
Ali Estilai
P. I. Forster
Shahina A. Ghazanfar
I. Grundy
Charlotte Gyllenhaal
Timothy Johns
Elaine Joyal
S. Kativu
David Lentz
Martin Luckert
Stephen P. McLaughlin
A. Mandondo
Betty J. Meggars
Brien Meilleur
Laura Merrick
D. W. Minter
Dan Moermann
Brad Morris
John M. Riddle
Marlene de Rios
Guillermo Schmeda Hirschmann
Douglas T. Seidel
Gerald J. Seiler
D. D. Soejarto
Jack Staub
Rexford H. Talbert
John Thieret
Sue Thompson
Arthur O. Tucker
Ghate Utkarsh
Susan Verhoek
High Wilson
E. T. F. Witkowski
Thomas M. Zennie
Karl Zimmerer

VOLUME 52: INDEX TO BOOK REVIEWS

VOLUME 52: INDEX TO BOOK REVIEWERS

Economic Botany Policies and Manuscript Requirements

Publication in *Economic Botany,* the journal of the Society for Economic Botany, is open to those who are members in good standing of the Society. When a paper has two or more authors, the author submitting the paper is expected to be a member. A cover letter accompanying a manuscript submission should state which author is responsible for revision and for reading proofs and that the manuscript is not under consideration by any other publication. Please also indicate the number of text pages, the number of tables, figures, and appendices.

Economic Botany specializes in scientific articles on the botany, history, and evolution of useful plants and their modes of use. Primarily agronomic or horticultural papers and those concerned mainly with analytical data on the chemical constituents of plants should be submitted elsewhere.

Papers to be considered for publication should be sent, in triplicate, to the editor. Send three prints for each photograph and an original and two clear copies of line drawings or graphs. Ordinarily, manuscripts should not exceed 30 double-spaced pages (ca. 7500 words) but longer manuscripts will be considered. In the letter accompanying the manuscript, please include the telephone number where the author can most readily be reached, and include fax numbers and e-mail addresses if possible.

All papers are peer reviewed; authors are welcome to include the names and addresses of objective reviewers to be contacted by the editor at the editor's discretion.

The *Council of Biological Editors Style Manual 5th ed.* is the primary reference used in setting the standards for the preparation of *Economic Botany* publications. The format and style of manuscripts (especially Literature Cited entries), further, must conform to the practices followed in the most recent issues of *Economic Botany. The Chicago Manual of Style*, 13th or 14th editions, The University of Chicago Press, Chicago and London, may be consulted for additional guidance.

Format. The manuscript should be typed or processed (10 or 12 characters/inch) on one side of the paper only and should be **double spaced throughout** (i.e., one **full** line of space between each two lines of text, tables, quotations, literature cited, **everywhere**). Use good quality paper ca. 21.5 × 28 cm (8.5 × 11 in), with **margins** (at least 1 inch) all around. **Right margins** should **not** be justified. Number pages consecutively in the upper right. **Underline nothing** except words to be set in italics. Page 1 title, author(s) and abstract format as it appears in recent issues should be carefully followed. Authors are encouraged to send electronic diskette copies with their original submission but especially with their revised manuscript copies.

Style. In the text, spell out one-digit numbers unless they are used with units of measure (four oranges, 4 cm) and use numerals for larger numbers. Do not begin sentences with an abbreviation or numeral. Do not use footnotes except for title and within tables.

Abstract. This should not be more than 150 words, in English, at the beginning of the article. A translation of the English abstract, in French, German, Portuguese, Russian, Spanish, or other appropriate language, with title, should follow the English abstract. Abstracts should be underlined **except** for plant names.

Authors' names for binomials should be included the first time the binomial appears in the text. **Cultivar** names are not italicized; they are either enclosed in single quotes (*Zea mays* 'Silver Queen') or preceded by "cv." (*Zea mays* cv. Silver Queen).

References to plants other than the most widely recognized species must be documented by reference to herbarium vouchers. Follow standard practice, i.e., use the collector's name(s), the collection number, and the acronym for the herbarium of deposit following *Index Herbariorum,* ed. 8 (e.g., *Smith 15467* TEX). Even well-known plants should be documented by herbarium vouchers or propagule sources in studies of infraspecific variation or similar research.

Authors not fluent in English should have their paper read by a botanical colleague who is fluent in English. Non-English abstracts should be read and corrected by a botanical colleague who is fluent in that language.

References in the text are to be cited by author(s) and date, e.g., a journal article (Rashford 1995); a book (Lewis and Elvin-Lewis 1977: 434); multiple citations should be in **alphabetical order** (e.g., Adams 1987; Martin 1922; Roberts 1949; Zimmermann 1813); multiple citations of the same author, in chronological order. For publications by 1–3 authors, name each author; more than three authors, use (first author, et al. date). **Unpublished references** should be used only if a reader, with reasonable effort, can obtain a copy. Reference to manuscripts which have not yet been published but which have been **accepted** for publication are "in press"). Use "n.d." for the date. **Journal titles should not be abbreviated.** In Literature Cited, underline nothing except Latin

Economic Botany 52(4) pp. 459–460. 1998

names, and for a book citation, "Pages xxx–xxx *in*...". Be certain that each item in the Literature Cited is cited in the text—and vice versa—and that spellings and dates are in agreement between Literature Cited and text. Double-check spellings and diacritical marks of names and titles; verify dates, volume numbers, and inclusive pagination. Papers with inaccuracies in bibliographic data are not acceptable.

Typical citations

Journal article:

Johns, T., and E. K. Kimanani. 1990. Herbal remedies of the Luo of Siaya district, Kenya: establishing quantitative criteria for consensus. Economic Botany 44:369–381.

Book:

Chapman, V. J., and D. J. Chapman. 1980. Seaweeds and their uses. 3rd ed. Chapman and Hall, London.

Patino, V. M. 1964. Plantas cultivadas y animales domesticos en America equinoccial. Vol. 2. Plantas alimenticias. Imprenta Departamental, Cali, Colombia.

Vavilov, N. I. 1992. Origin and geography of cultivated plants. Ed. V. F. Dorofeyev; translated, Doris Löve, Cambridge University Press, Cambridge.

Book reprinted:

Millspaugh, C. F. 1974. American medicinal plants. Dover Publications, New York. Reprint of a work first published, as *Medicinal plants*, by John C. Yorston & Co., Philadelphia, in 1892.

Part of a book:

Zohary, Daniel. 1989. Domestication of the Southwest Asian Neolithic crop assemblage of cereals, pulses, and flax: the evidence from living plants. Pages 358–373 *in* David R. Harris, and Gordon C. Hillman, eds. Foraging and Farming, the Evolution of Plant Exploitation. Unwin Hyman, London.

Figures (line drawings, maps, graphs, photographs) intended for publication in *Economic Botany* must be of high quality and designed to make effective use of the space made available by the two column format. Photographs should be black and white, 5 × 7″ is recommended. Color prints may be submitted for black and white reproduction. The cost of **color reproduction** is high and must be borne by the author—consult the editor before submitting color prints. Most figures will not occupy a full page, they will be reduced to the width of a single column (65 mm) or two columns (140 mm), authors should be certain that lettering will be legible after reduction.

Single photographs should be loosely mounted on stiff paper and labeled on the back with the figure number and name of the author. Photographs **grouped together as a plate** must be accompanied by a sheet bearing the full caption and must be mounted on stiff white board with no white space between them (this is added by the press). Regardless of how many photographs comprise a plate, the plate itself must be rectangular in outline. Each illustration in a plate should be numbered as a separate figure—e.g., if the first four photographs of the manuscript are grouped together, a number is applied to each and the caption should read "Fig. 1–4." Numbers and lettering on illustrations must be done with a lettering instrument, press-on letters, or laser printer. Cover each Figure or plate with a protective sheet of paper, attach a separate 8.5 × 11 inch sheet bearing legend or caption, and number the Figures consecutively throughout the text. Avoid submitting illustrations (or anything else) on sheets larger than 8.5 × 11 inches.

The amount of tabular and illustrative material must be in reasonable proportion to the length of the text. Indicate, in the left-hand margin of the paper, where figures and tables are to be placed (e.g., "Fig. 1 [or Table 1] here").

Notes on economic plants submitted to *Economic Botany* should concern one species or a small group of species related by taxonomy or by use. Each note should be no more than five typed pages and should conform precisely to the format of notes already published. Illustrations, if any, should be designed to occupy no more than one printed journal page.

Page charges. All authors will be asked to pay page charges. Such payment, however, is not a condition of acceptance of papers.

Address correspondence and send manuscripts to Dr. Lawrence Kaplan, Editor, *Economic Botany,* Department of Biology, University of Massachusetts/Boston, 100 Morrissey Blvd., Boston, MA 02125-3393 U.S.A. Phone (617) 287-6606 or 287-6600; Fax (617) 287-6650; e-mail "lkaplan@umbsky.cc.umb.edu.

ECONOMIC BOTANY

Devoted to Past, Present, and Future Uses of Plants by People

Founded by

Edmund H. Fulling

Publication of The Society for Economic Botany

VOLUME 52

1998

Published for The Society

by

THE NEW YORK BOTANICAL GARDEN PRESS

Printed by
Allen Press, Inc.
Lawrence, Kansas

UNITED STATES POSTAL SERVICE™

Statement of Ownership, Management, and Circulation

(Required by 39 USC 3685)

1. Publication Title Economic Botany	2. Publication Number 0 0 1 3 - 0 0 0 1	3. Filing Date 9/18/98
4. Issue Frequency Quarterly	5. Number of Issues Published Annually 4	6. Annual Subscription Price $88.00
7. Complete Mailing Address of Known Office of Publication *(Not printer) (Street, city, county, state, and ZIP+4)* The New York Botanical Garden Press The New York Botanical Garden Bronx, NY 10458-5126		Contact Person Susan Frayman Telephone (718) 817-8957

8. Complete Mailing Address of Headquarters or General Business Office of Publisher *(Not printer)*

The New York Botanical Garden Press
The New York Botanical Garden
Bronx, NY 10458-5126

9. Full Names and Complete Mailing Addresses of Publisher, Editor, and Managing Editor *(Do not leave blank)*

Publisher *(Name and complete mailing address)*

The New York Botanical Garden Press
The New York Botanical Garden
Bronx, NY 10458-5126

Editor *(Name and complete mailing address)*

Dr. Lawrence Kaplan
Department of Biology, University of Massachusetts/Boston
Boston, MA 02125

Managing Editor *(Name and complete mailing address)*

Susan Frayman
The New York Botanical Garden Press
The New York Botanical Garden, Bronx, NY 10458-5126

10. Owner *(Do not leave blank. If the publication is owned by a corporation, give the name and address of the corporation immediately followed by the names and addresses of all stockholders owning or holding 1 percent or more of the total amount of stock. If not owned by a corporation, give the names and addresses of the individual owners. If owned by a partnership or other unincorporated firm, give its name and address as well as those of each individual owner. If the publication is published by a nonprofit organization, give its name and address.)*

Full Name	Complete Mailing Address
The New York Botanical Garden	Bronx, NY 10458-5126

11. Known Bondholders, Mortgagees, and Other Security Holders Owning or Holding 1 Percent or More of Total Amount of Bonds, Mortgages, or Other Securities. If none, check box → ☒ None

Full Name	Complete Mailing Address

12. Tax Status *(For completion by nonprofit organizations authorized to mail at special rates) (Check one)*
The purpose, function, and nonprofit status of this organization and the exempt status for federal income tax purposes:
☒ Has Not Changed During Preceding 12 Months
☐ Has Changed During Preceding 12 Months *(Publisher must submit explanation of change with this statement)*

PS Form **3526**, September 1995 *(See Instructions on Reverse)*

13. Publication Title	14. Issue Date for Circulation Data Below
Economic Botany	Volume 52(3), 8/21/98

15. Extent and Nature of Circulation		Average No. Copies Each Issue During Preceding 12 Months	Actual No. Copies of Single Issue Published Nearest to Filing Date
a. Total Number of Copies *(Net press run)*		2,200	2,200
b. Paid and/or Requested Circulation	(1) Sales Through Dealers and Carriers, Street Vendors, and Counter Sales *(Not mailed)*	None	None
	(2) Paid or Requested Mail Subscriptions *(Include advertiser's proof copies and exchange copies)*	1,869	1,914
c. Total Paid and/or Requested Circulation *(Sum of 15b(1) and 15b(2))* ▶		1,869	1,914
d. Free Distribution by Mail *(Samples, complimentary, and other free)*		6	7
e. Free Distribution Outside the Mail *(Carriers or other means)*		None	None
f. Total Free Distribution *(Sum of 15d and 15e)* ▶		6	7
g. Total Distribution *(Sum of 15c and 15f)* ▶		1,875	1,921
h. Copies not Distributed	(1) Office Use, Leftovers, Spoiled	325	279
	(2) Returns from News Agents	None	None
i. Total *(Sum of 15g, 15h(1), and 15h(2))* ▶		2,200	2,200
Percent Paid and/or Requested Circulation *(15c / 15g x 100)*		99.7%	99.6%

16. Publication of Statement of Ownership
☒ Publication required. Will be printed in the 52(4) issue of this publication.
☐ Publication not required.

17. Signature and Title of Editor, Publisher, Business Manager, or Owner	Date
Susan Frayman, Director, The New York Botanical Garden Press *Susan Frayman*	9/18/98

I certify that all information furnished on this form is true and complete. I understand that anyone who furnishes false or misleading information on this form or who omits material or information requested on the form may be subject to criminal sanctions (including fines and imprisonment) and/or civil sanctions (including multiple damages and civil penalties).

Instructions to Publishers

1. Complete and file one copy of this form with your postmaster annually on or before October 1. Keep a copy of the completed form for your records.
2. In cases where the stockholder or security holder is a trustee, include in items 10 and 11 the name of the person or corporation for whom the trustee is acting. Also include the names and addresses of individuals who are stockholders who own or hold 1 percent or more of the total amount of bonds, mortgages, or other securities of the publishing corporation. In item 11, if none, check the box. Use blank sheets if more space is required.
3. Be sure to furnish all circulation information called for in item 15. Free circulation must be shown in items 15d, e, and f.
4. If the publication had second-class authorization as a general or requester publication, this Statement of Ownership, Management, and Circulation must be published; it must be printed in any issue in October or, if the publication is not published during October, the first issue printed after October.
5. In item 16, indicate the date of the issue in which this Statement of Ownership will be published.
6. Item 17 must be signed.

Failure to file or publish a statement of ownership may lead to suspension of second-class authorization.

PS Form **3526**, September 1995 *(Reverse)*

NUMBER 1
January–March 1998

Issued 31 March 1998

NUMBER 2
April–June 1998

Issued 16 June 1998

NUMBER 3
July–September 1998

Issued 21 August 1998

NUMBER 4
October–December 1998

Issued 30 November 1998